1988

The Merging of Disciplines:
New Directions in Pure, Applied,
and Computational Mathematics

The Merging of Disciplines:
New Directions in Pure, Applied, and Computational Mathematics

Proceedings of a Symposium
Held in Honor of Gail S. Young
at the University of Wyoming, August 8–10, 1985.
Sponsored by the Sloan Foundation, the National Science Foundation, and
the Air Force Office of Scientific Research

Edited by
Richard E. Ewing
Kenneth I. Gross
Clyde F. Martin

With 25 Illustrations

Springer-Verlag
New York Berlin Heidelberg
London Paris Tokyo

Richard E. Ewing
Department of Mathematics
University of Wyoming
Laramie, WY 82071, USA

Clyde F. Martin
Department of Mathematics
Texas Tech University
Lubbock, TX 79409, USA

Kenneth I. Gross
Department of Mathematics
University of Wyoming
Laramie, WY 82071, USA

Library of Congress Cataloging in Publication Data
Symposium on New Directions in Applied and Computa-
  tional Mathematics (1985 : University of Wyoming)
  The merging of disciplines.
  "Proceedings of the Symposium on New Directions
in Applied and Computational Mathematics."
  1. Mathematics—Congresses. 2. Young, Gail S.
I. Ewing, Richard E. II. Gross, Kenneth I.
III. Martin, Clyde. IV. Title.
QA1.S8993 1985   510   86-24848

Printed and bound by R.R. Donnelley & Sons, Harrisonburg, Virginia.
Printed in the United States of America.

9 8 7 6 5 4 3 2 1

ISBN 0-387-96414-2 Springer-Verlag New York Berlin Heidelberg
ISBN 3-540-96414-2 Springer-Verlag Berlin Heidelberg New York

Dedicated to the memory of
Irene V. Young
October 6, 1925 – October 4, 1985

GAIL S. YOUNG

PREFACE

This volume is the Proceedings of the symposium held at the University of Wyoming in August, 1985, to honor Gail Young on his seventieth birthday (which actually took place on October 3, 1985) and on the occasion of his retirement.

Nothing can seem more natural to a mathematician in this country than to honor Gail Young. Gail embodies all the qualities that a mathematician should possess. He is an active and effective research mathematician, having written over sixty papers in topology, n-dimensional analysis, complex variables, and "miscellanea." He is an outstanding expositor, as his fine book *Topology*, written with J. G. Hocking (Addison Wesley, 1961), amply demonstrates. He has a superlative record in public office of outstanding, unstinting service to the mathematical community and to the cause of education. But what makes Gail unique and special is that throughout all aspects of his distinguished career, he has emphasized human values in everything he has done. In touching the lives of so many of us, he has advanced the entire profession. Deservedly, he has innumerable friends in the mathematical community, the academic community, and beyond.

It is tempting to describe Gail Young today as the archetypal "elder statesman" of mathematics; but that description misses the point that Gail never strikes one as an elder! He is far too vigorous, too enthusiastic, too involved to qualify for any title suggesting withdrawal and dispassionate reflection. Our respect for Gail certainly includes an appreciation of his wisdom; but we have appreciated his wisdom for decades. He has not had to wait till now to achieve it, nor have we waited till now to evidence our recognition of it.

It is appropriate to list here certain significant details of Gail's life and career. But such a recital, impressive though it may be, is still quite inadequate to capture the flavor and the true meaning of Gail's role in American—and world—mathematics over the past thirty years. The whole man whom we are privileged to know, to respect and to love is not to be found in the details about to be presented. Gail Young was born in Chicago, Illinois, on October 3, 1915. He went as an undergraduate to Tulane University in 1935 and transferred to the University of Texas for his senior year. While at Tulane he was profoundly influenced by H. E. Buchanan, who started many mathematicians on their careers, and by Bill Duren, to whom Gail gives credit for introducing him to rigorous mathematics. Gail remained at Texas to do his Ph.D. under R. L. Moore, obtaining the degree in 1942. He taught at Purdue University until 1947 and then at the University of Michigan through 1959, attaining the rank of full professor. He then returned to Tulane University where he stayed until 1970, serving as chairman for the years 1963–68. He was at Rochester University from 1970 to 1979, most of the time as chairman, moved to Case Western Reserve University in 1979 and to a visiting position at the University of Wyoming in 1981, where he has remained as vigorous, active, and effective as ever.

Gail's services to the mathematical community and to the cause of education through his work within various professional organizations, are too numerous to list individually. Of particular significance are his terms as President of the MAA (1969–70); his chairmanship of the Teacher Training Panel of CUPM (1964–68); his membership of the AMS Committee on Employment and Educational Policy (1972–74) and the Committee on Women in Mathematics (1974–76); his membership of the Council of AAAS (1968–70, 1975) and his chairmanship of the mathematics section (1981–84); and his membership of the NAS-NRC Committee on Applications of Mathematics (1964–67). Yet this catalog of outstanding service omits so much which most of us would be proud to claim as our own principal contributions.

The simple truth is that everything Gail has done has been well done, useful and important. And he has done an enormous lot! Let me, then, desist from the unwonted objectivity of this listing of Gail's work as researcher, teacher and campaigner for better education, and adopt the frankly subjective mode appropriate to a friend and admirer.

Gail was, by all accounts, an outstanding student. The story is told of Gail returning to calculus class after an absence of several weeks (that was during the Depression and Gail had been unable to pay the tuition fees!). On his first day back there was a test from which Buchanan would gladly have excused Gail. Gail insisted on taking the test and obtained the grade of 100, though others found the test awesomely difficult. Buchanan, reporting Gail's success and the failure of others whom he did not name, asked "What does that show?" After a short silence, one student replied, "I beg your pardon, Dr. Buck, but it shows that you ruined us."

Many of Gail's students have testified to his remarkable qualities as a teacher, at all levels. He did not use the "Moore" method of his own teacher, but neither was he a lecturer in the traditional mold. He showed his students how to do mathematics—how to pose questions and to try to answer them, how to develop and exploit geometric insight. In his view of mathematics as consisting of the search for questions it might be profitable to try to solve, he much resembles the great Swiss topologist Heinz Hopf. Moreover, he shares with Hopf the ineffable quality of unbounded kindness towards all genuinely interested in mathematics and trying to learn. As one of his students has said, "When he was with one of his students, he was 100% with that student... He knew the state of mind of his students, and could be encouraging when things weren't going well mathematically or personally."

My own association with Gail began in 1964 when he persuaded me to join the Teacher Training Panel of CUPM. I joined that panel knowing little of its duties because I was aware of wanting to work with Gail—just as I had started to study algebraic topology in 1946, knowing little of the subject, because I wanted to work with Henry Whitehead. Both men possessed to a remarkable degree the ability to convey the conviction—fully vindicated by subsequent experience—that to work with them would be to do something which was important, probably hard, but certainly fun. Throughout my many associations with Gail, I have always been struck by the sheer massive good sense of the man. Gail is possessed of a universal intelligence and sensitivity which are by no means guaranteed by possession of the quality of cleverness. Testimony to that good sense is provided by the anecdote told by Henry Pollak relating to their visit to Africa in 1968 to evaluate the effects of the Africa Mathematics Program (usually called the Entebbe Program). In Addis Ababa, Gail and Henry had an interview with a high official of the Ethiopian Ministry of Education. At the end, Gail quietly suggested to Henry that he be left alone for a few minutes with the official. It transpired that the official was making a grant application to the Ford Foundation and required some help in drafting the proposal. The official realized that Gail was the person to provide that help—and Gail understood the loss of face that would be involved if more than he and the official were present.

Further testimony to Gail's good sense, and to his total honesty and great integrity, is provided by the record of his chairmanship of the Tulane department in the years 1963–68. For various evident reasons this was not a propitious time to try to build a strong research department at a university in the South, but Gail was remarkably successful. Ed Dubinsky refers to his "openness, reasonableness and concern," and has described how Gail wooed him to Tulane by writing him a letter in which he described his philosophy in running a department, offered a salary which he acknowledged to be too low, and then explained how it was arrived at. One suspects that it needed a combination of Gail's honesty and Ed's idealism to have created this match!

No tribute to Gail would be complete without mention of his wonderful wife Irene, who was the Administrative Assistant of the Tulane department. Theirs was an ideal marriage, a true partnership. Our sympathies go out to Gail who must face his retirement deprived of her loving companionship; but we find some comfort in recalling her pleasure in observing the outpouring of respect, admiration, and affection for Gail at the banquet in his honor held in conjunction with the Laramie Conference.

My final word should be this: Gail, though now retired, is still an active member of our community and our profession—long may it be so, for we need him for his wisdom and we enjoy him for himself.

PETER HILTON

with the assistance of

Ray Cannon, Ed Dubinsky, Bill Duren, Ken Gross, J. G. Hocking,

Burton Jones, Henry Pollak and Sanford Segal

CONTENTS

SYMPOSIUM

on

NEW DIRECTIONS IN APPLIED
AND COMPUTATIONAL MATHEMATICS

August 8–10, 1985

Thursday, August 8

Opening remarks by Dr. Donald Veal
President of the University of Wyoming

First Session

HENRY O. POLLAK, AT&T Communications
"Pure and applied mathematics from an industrial perspective"

CHRISTOPHER BYRNES, Arizona State University
"Recent results in adaptive control: An opportunity for intelligent and artificially intelligent control"

Second Session

ANIL NERODE, Cornell University
"The impact of logic and recursive functions on computational sciences"

ISMAEL HERRERA, Universidad Nacional Autonoma de Mexico
"Some unifying concepts in applied mathematics"

First Session for Contributed Papers

HARLEY FLANDERS, Florida Atlantic University
"Symbolic manipulation"

TETSURO YAMAMOTO, University of Wisconsin-Madison
"Error bounds for Newton's method under the Kantorovich assumptions"

KENNETH J. HOCHBERG, Case Western Reserve University
"New directions in mathematical population biology"

Friday, August 9

Third Session

PETER J. HILTON, State University of New York at Binghamton
"Teaching and research - the history of a pseudo-conflict"

CLYDE F. MARTIN, Texas Tech University
"Observability in dynamical systems"

Fourth Session

JACOB T. SCHWARTZ, Courant Institute
"Problems of shape recognition"

DANIEL J. KLEITMAN, MIT
"Combinatorics and applications, a mutual enrichment"

Second Session for Contributed Papers

DONALD RICHARDS, University of North Carolina
"Letter values in multivariate exploratory data analysis"

FRANK HARARY, University of Colorado, Boulder
"Graph theoretic models in anthropology, biology and chemistry"

G. ELTON GRAVES, Rose-Hulman Institute of Technology
"Computer graphics in numerical analysis and differential equations courses"

Saturday, August 10

Fifth Session

RICHARD E. EWING, University of Wyoming
"Mathematical modeling and large-scale computing in energy and environmental research"

THOMAS F. BANCHOFF, Brown University
"Computer graphics applications in geometry: 'because the light is better over here'"

Sixth Session

STEPHEN SMALE, University of California at Berkeley
"When and how fast can Newton's method be expected to converge?"

Panel Discussion
"Implications for undergraduate and graduate education in mathematics"
WILLIAM L. DUREN, JR., University of Virginia
SOL GARFUNKEL, COMAP
PETER J. HILTON, SUNY at Binghamton
GAIL S. YOUNG, University of Wyoming
KENNETH I. GROSS, University of Wyoming (moderator)

Banquet in honor of Gail Young

INTRODUCTION

The articles that follow form the Proceedings of a truly remarkable symposium, held at the University of Wyoming in August, 1985, on the theme "New Directions in Applied and Computational Mathematics." The result was successful beyond anyone's expectation. The spirit of communication, enthusiasm, and cooperation was pervasive, and the breadth and depth of topics was spectacular. All who attended came away with a rich and enjoyable experience. We hope that the reader of this Proceedings will also.

The focus of the Symposium was on the mutual interaction among pure mathematics, applied mathematics, and computer science that is rapidly and dramatically changing the nature of all three disciplines. Indeed, when the editors were in college and graduate school in the late fifties, the sixties, and the early seventies, there was seldom a difficulty in determining who were the applied mathematicians and who the pure. Now, even the purest of mathematics has found profound and influential applications, and applied problems have generated new thrusts in areas that have always been regarded as pure. The main instrument in this coalescence has been the computer.

A glance at the list of authors and titles reveals the unique flavor of the Symposium and this volume. A number of distinguished mathematicians, whose careers and whose research illustrate the unity of mathematics, were asked to share their insights into this phenomenon and to describe their own contributions. Thus, the articles which appear in this volume, which also includes several contributed papers, are diverse in both scope and nature. Some are expositions of the cultural context of a field of mathematics, a concept, or a perspective. Others present important current research. Their quality and clarity is exceptional. At the very least, the articles that appear here should dispel forever the notion that there are two kinds of mathematics, and they should reinforce the idea that it is impossible to predict the source of the next breakthrough. The labels "pure" and "applied" are no longer applicable to mathematics or to mathematicians.

It is most appropriate that this volume is in honor of Gail Young, who long ago realized that mathematics should not be separated from its applications. That is one of many examples of his wisdom, from which all three of us have benefitted.

It should be noted that the current Symposium is a sequel to an earlier one in 1981, organized by Gail and Peter Hilton, with a similar title, "New Directions in Applied Mathematics." That conference also had as its theme the rapport of pure and applied mathematics. It emphasized the fact that modern techniques are of critical importance in applications. Missing from that conference was the computer, as the organizers were not yet ready to define the role of the computer in its interaction with mathematics and applications. To fill that gap was a motivation of this current Symposium.

As we learn from the articles herein, much has transpired since the earlier conference. No doubt the same statement will be applicable to this volume a few years hence. The rapid developments alluded to at the outset of this introduction will continue to bring together more areas that were previously thought unrelated. That can be seen already. For example, who would have predicted at the time this symposium was organized that the von Neumann algebras which arose decades ago in quantum physics and had just recently become a major tool in knot classification, would—because of the resultant improvements in classification—excite great interest among microbiologists concerned with the knotting of DNA? This would be a major topic were the Symposium held today. Thus, the editors see a clear need to continue this series of symposia periodically as new developments dictate.

To close on a personal note, it is a pleasure to express our appreciation to all who helped make the Symposium and this Proceedings a success. We are deeply indebted to the speakers, authors, and participants. Special thanks go to a number of individuals. Sol Garfunkel played a major role in proposal preparation. The administrative assistance of Sharon Distance in organizing the Symposium, and Lois Kline and Paula Sircin in preparation of the Proceedings was invaluable. In particular, the excellence of Paula's typing is evidenced in the camera ready copy that is before you. The support, encouragement, and suggestions by Dr. Walter Kaufmann-Buhler, Editor for SpringerVerlag, have been of great benefit to us. Finally, none of this would have been possible without the financial support of the sponsors. The Editors wish to record their gratitude for the generosity of the Sloan Foundation, the National Science Foundation, and the Air Force Office of Scientific Research.

RICHARD E. EWING
KENNETH I. GROSS
CLYDE F. MARTIN
Laramie, Wyoming
May 25, 1986

COMPUTER GRAPHICS APPLICATIONS IN GEOMETRY: "BECAUSE THE LIGHT IS BETTER OVER HERE"

THOMAS F. BANCHOFF

Department of Mathematics
Brown University
Providence, Rhode Island 02912

The theme *New Directions in Applied and Computational Mathematics* gives all of us who think of ourselves as "pure mathematicians" the chance to reexamine the changes that have taken place in our understanding of that term as a result of the new directions which have appeared in our lifetimes. I have always thought of myself as a geometer, as soon as I realized that it was possible to think of different subspecies of mathematician, and I can make the case today that there will be major differences in the way we do geometry and the way we present our insights to our students and colleagues because of the dramatic developments in visualization capabilities in the form of interactive computer graphics. One effect of these developments is that I find myself revising my concept of the differences between pure and applied mathematics. A virtue of a symposium such as this, honoring a wide-ranging mathematician like Gail Young, is that we speakers are encouraged to be introspective about these changes in our careers, and of course that encourages in turn a certain anecdotal style.

When I was an undergraduate student at Notre Dame I had the chance to form definite attitudes about the nature of applied mathematicians as I watched with some fascination and dismay as my sophomore roommate gradually became one. Whereas I took philosophy and literature courses to round out my abstract mathematics, he spent time in the *physics* and chemistry *labs*. He actually read the optional *applications* chapters on fluid flow when we took an abstract graduate course in complex analysis. He used a slide rule (which, for the benefit of the young people in the audience, was a wooden analogue *calculating device* attached to the belt).

We both ended up as mathematics graduate students at Berkeley in 1960 and our differences became even greater. Whereas I took the geometry and topology option, he chose to study *differential equations*. He began to spend more and more time with

1

numerical computations using *computers*, and he would rail against the evils of bugs and batch processing. Ultimately he moved over the line into theoretical physics, where he worked in a *laboratory* on *other people's problems*. He wrote *joint papers* with *federal funding*. While I went back to Notre Dame to teach in Arnold Ross's summer program, he worked as a *consultant* and he began to *make money*. All these italicized characteristics I decided were the marks of an applied mathematician. Little did I suspect that virtually all of them would gradually begin to describe the work of pure mathematicians as well, precisely under the influence of several of the new directions in applied and computational mathematics that we have been examining here as part of this symposium.

If applied mathematicians work on other people's problems, I soon found out where pure mathematics graduate students got their problems—from their thesis advisors. Professor Chern (thesis advisors often retain formal titles forever) suggested that I study a recent paper of Louis Nirenberg "On a class of closed surfaces" from a symposium on differential equations (!) at Wisconsin. I had to scurry to learn topics I thought only those other people had to know, like Poincaré-Bendixson theorems and existence and uniqueness results for hyperbolic partial differential equations. But true to an earlier "pure" geometric impulse, I tried hard to recast the definitions used by Nirenberg and twenty-five years previously by A. D. Alexandroff in a way that would apply equally well to polyhedra. The result was a surprising simplification of the notion of minimal total absolute curvature which ended up applying to an even larger class of surfaces.

Alexandroff restricted himself to real analytic surfaces and he studied embeddings for which the total absolute curvature was as small as it could be. For surfaces homeomorphic to a sphere, this condition is equivalent to convexity, and for surfaces homeomorphic to a torus, the corresponding condition is that all the positive curvature was concentrated on the "outer part," the intersection of the surface with the boundary of its convex hull. Alexandroff called these objects "*T*-surfaces" and he was able to show that they shared some of the rigidity properties of convex surfaces. In particular, any two isometric *T*-surfaces had to be congruent so that any one-to-one mapping between such surfaces which preserved intrinsic distances could be extended to an isometry of all of Euclidean 3-space. Nirenberg in 1963 proved analogous theorems for smooth surfaces which satisfied some additional technical hypotheses necessitated by his differential equations techniques. He used a definition equivalent to Alexandroff's which said that any local support tangent plane to the surface, intersecting the surface locally at one point, must be a global support plane, meeting the surface at no further point. I observed that this definition was sufficiently geometric to apply to polyhedral surfaces as well, so I determined to remove the extra hypotheses from Nirenberg's theorem by proving a polyhedral rigidity theorem and applying approximation techniques such as had been used successfully by Pogoreloff and his school in the case of convex surfaces. (I later found that George Orland and J. J. Stoker were working on similar ideas.)

Unfortunately the approach failed completely. After a number of months of trying to prove the result, I found a counterexample, a pair of isometric polyhedral tori such that every local support plane is a global support plane and the surfaces are not congruent even though they are built from congruent polygonal faces with the same identifications. Thus if it is possible to eliminate some of the technical hypotheses in Nirenberg's rigidity theorem for smooth surfaces, it will have to be done in some other way. To my knowledge, this question is still open.

The polyhedral non-rigidity examples did ultimately become a part of my thesis, and they inaugurated for me a theme that has characterized a good deal of my research over the years: Start with a problem in the theory of smooth surfaces and identify the geometric content so that the question makes sense for polyhedra as well. Then either prove the result to establish an analogue of the smooth theorem, or disprove it to show how the smoothness hypothesis is used in an essential way.

This program took hold firmly when Professor Chern introduced me to a visiting geometer who he said was interested in the same kinds of problems, Nicolaas Kuiper. When he saw my models, Kuiper suggested that I might try to prove some analogues of his theorems about surfaces with minimal total absolute curvature in higher dimensional spaces. Chern and Lashof had observed that the theory of total absolute curvature could be recast in the language of critical point theory, and Kuiper had exploited this fact in the case of smooth surfaces to show that if a smooth surface was embedded in an n-dimensional Euclidean space so that no height function had more than one local maximum, then the surface had to be contained in a 5-dimensional affine subspace, and moreover, if it did not lie in a 4-dimensional space, then it had to be a very specific surface, the Veronese embedding of the real projective plane. I set out to establish the analogue of this result in the case of polyhedral surfaces in higher dimensional space (I was quite fond of the geometry of higher dimensional space but I had never expected that my intuition there would play much of a role in actual research.) The project failed in a rather spectacular fashion. Within a few weeks I had found a whole family of polyhedra, one for each dimension, which satisfied the condition of having the minimal number of critical points in every direction but which were not contained in any hyperplane. Kuiper was surprised. He looked at me and said, "I'll give you six months. If you haven't written your thesis by that time, I'm going to give the problem to one of my students. It's too nice not to be done by someone."

Fortunately it didn't take me more than a few months to finish. Things in the polyhedral case were quite different than in the smooth case, but there was still a theory there, and one populated with new examples. I developed the first theorems in the subject and I had my thesis. It was quite far from the original "applied" theorems that started it out.

It may be of some historical interest to tell the origin of the name "tight" which has become established enough to have a place in the *Mathematical Reviews*. Chern felt that the concept of minimal total absolute curvature needed a name, and the terms in use, like "minimal" and "convex" already had other connotations. Since Alexandroff had used the term "*T*-surfaces" there was a strong bias for a word beginning with "t". I came up with "tight," "taut" (and my favorite) "turgid." Chern felt that the last didn't have the right sound, and "taut" was already in use in another part of mathematics. So "tight" it became, and my thesis was on "Tightly Embedded 2-Dimensional Polyhedral Manifolds."

On the way to my thesis results I discovered an alternative definition for tight surfaces that was so simple that at first I suppressed it, thinking that anyone who heard it would find the subject just too trivial to consider. The condition is the *two-piece property*. An object in a Euclidean space is said to have the two-piece property if any hyperplane cuts it into at most two pieces. An object in the plane therefore has the TPP if and only if every line separates it into at most two pieces. Already we can investigate a wealth of examples (using shapes suggested by the presentation of Jack Schwartz earlier in the symposium).

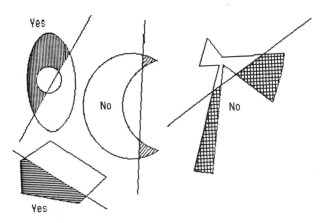

Figure 1

I was somewhat embarrassed by the fact that alone among my friends I was able to explain to strangers what my thesis was about—I could pick up a knife and demonstrate that some objects on the table could be cut into more than two pieces by a single straight cut and that others could not. In some sense it seemed more practical than K-theory or algebraic geometry but one had to be suspicious of mathematics that was too understandable.

The two-piece property turned out to be much more powerful than anyone imagined at first, and at the hands of William Pohl and John Little it provided a strong generalization of Kuiper's theorem about the smooth Veronese surface to arbitrary dimension. Even more surprising was a purely topological result of Kuiper and Pohl which stated that if the real projective plane is embedded in 5-space with the two-piece property, and not lying in any affine 4-space, then the embedding is either the analytic Veronese surface or the polyhedral embedding obtained by sending the six antipodal pairs of vertices of the icosahedron to six points in general position, as described in my thesis.

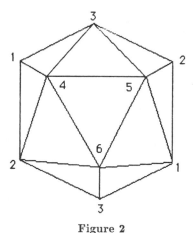

Figure 2

No other real projective planes with the TPP exist in 5-space. To my knowledge there are no other results of this type where such a weak topological condition leads to precisely two quite different solutions. One of the most interesting outstanding conjectures along these lines is that the same situation should be true for embeddings of the complex projective plane in 8-space since the only two known examples are the analytic Mannoury embedding analogous to the Veronese surface and the remarkable 9-vertex triangulation of CP^2 by Wolfgang Kühnel. (These embeddings are described in the article by Kühnel and the author cited in the bibliography.)

Up to this time, the only tools I used for studying these surfaces were drawings and cardboard models, but these were becoming more and more inadequate, especially for surfaces in 4- and 5-dimensional space. When I came to Brown however all of that changed when I was introduced to Charles Strauss. He was just completing his Ph.D. thesis in applied mathematics on interactive graphics techniques in problems of design of systems of pipes. His graphics programs did the work of eleven engineers but at the time they cost more to install and operate than the salaries of eleven engineers so the project became of "academic interest." Charles had a marvelous new technique and he needed new problems. I had great problems and I needed just the capabilities his program had. We found ourselves in a very unlikely pure-and-applied collaboration which lasted a dozen years. Neither had to learn the subject of the other, but once we learned to communicate, we could do things neither had suspected possible. We were able to produce images which were projections of surfaces in 4-space and to manipulate those projections so as to make direct visual investigations of phenomena we previously knew only through formulas and abstract arguments. It was an extremely exciting time, resulting in a film of TPP surfaces in the 3-sphere which changed our whole perception of the subject.

The study of tight surfaces in the 3-sphere in 4-space was even more embarrassing from the point of view of understandability. Any tight surface in the 3-sphere could be projected stereographically into 3-space to yield a surface not only with the TPP but also with the *spherical two-piece property*. An object has the STPP if any sphere cuts the object into at most two pieces. This property could be described even to nonmathematicians by showing that you could never cut a doughnut into more than two pieces by any one bite. When Robertson and Carter wanted to find a term for the generalization of the STPP to higher dimensions, they chose the word "taut"!

Our film showed a one-parameter family of conformally equivalent cyclides of Dupin obtained by rotating the 3-sphere and keeping track of the images of the product torus under stereographic projection. It took very long to generate each frame and it was only when we showed the film that we realized how much more powerful that medium could be than any collection of single frames. We *saw* an object from the fourth dimension, and we knew at that moment that we would never again be satisfied only with static images.

We realized only later how lucky we had been to come up with a usable, not to mention striking, film on our first try. It took us several years more before we were able to create something new that we could be satisfied with, and that was the study of perhaps the simplest of objects in the fourth dimension, the 4-dimensional cube or *hypercube*. Again for historical purposes it should be mentioned that the hypercube was already the subject of a stereoscopic film by A. Michael Noll at Bell Laboratories in the early 1960s. As it happens, he concluded that the project did not seem to give any new insight into the visualization of objects in 4-space. Perhaps the introduction of color (by means of filters) provides something new, since in any case the "grand tour of the hypercube" has become a primer for the understanding of any object in 4-space by means of projections into one-parameter families of 2-planes, in orthographic and in central projections. People see it dozens of times and continue to discover new phenomena, and the viewing experience paves the way for visual examination of more complicated objects in 4-space.

One of our favorite subjects is the Veronese surface, the tight algebraic embedding of the real projective plane into 5-space. One of the algebraic properties of this mapping is that almost any projection into a 4-dimensional subspace is an embedding, so we immediately can get the object into a dimension which is "familiar." We then project further into a one-parameter family of 3-dimensional hyperplanes to obtain a film view of the Veronese surface. Actually we developed parts of this film over a period of years so that it presents a history of the way changing computer graphics techniques have highlighted totally different aspects of the subject.

When we first studied surfaces, all of our objects were described in "wire frame mode," where the surface was presented in the form of the 1-skeleton of a polyhedral approximation of the object obtained by the images of grid lines in coordinate charts. Such objects are necessarily transparent, and the primary features which lend themselves to mathematical exploration are the folds and cusps of the projected image. The power of this technique first became apparent to me in pre-computer days when I was teaching at Harvard and as I walked past the Carpenter Center for Visual Arts, I saw the spinning forms of wire-screen models suspended from threads in the second floor window. The way that folds and cusps were created and annihilated as the object rotated was fascinating. I determined then and there that someday I would study the geometry of those transformations. This vow was fulfilled years later in the study of singularities of Gauss mappings together with Clint McCrory and Terence Gaffney in our geometry seminar at Brown in which all the key examples were identified and presented using computer graphics images.

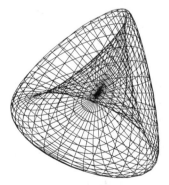

Figure 3. The Veronese Surface, wire frame mode.

The use of film to present geometric properties of TPP surfaces became more persuasive with the introduction of color, so that we could show the geometric structure of level sets of height functions on surfaces by changing the color of the wire-frame object as the level rose, especially in passing a critical level. Once again a new technique had caused a shift in our research directions and we concentrated for a time on the Morse theory of surfaces. We were very pleased when one of the visitors to our laboratory was Marston Morse himself.

A crisis occurred when simultaneously our locally devised vector graphics equipment reached a state of unrepairability and Charles Strauss became so busy as a consultant on mortgage and leasing programs that he no longer had time for academic computer graphics. I wondered if I was doomed to return to the world of cardboard and tape. But no, in the meantime there had grown up a cadre of students who were so familiar with the machinery as well as the programming that they were ready to step in and assume a full role in the collaborative mode that I had learned to enjoy. And at the same time there appeared powerful new machines which were able to realize some of a geometer's dream. The raster graphics technology could produce filled-in images with shading and highlighting, rendering these abstract objects as if they had been fashioned in gleaming metal. The student who developed most of the techniques we used during this period was David Salesin, who left Brown a semester early to join the Lucasfilm imaging team to bring his computer graphics talents to the creation of different sorts of special effects.

As the techniques shifted, so did the mathematical questions that were most natural to ask. Instead of investigating folds and cusps of projections to planes, we now considered the double points and triple points of immersions of surfaces and the "pinch point" singularities of generic projections of surfaces from 4-space into 3-dimensional spaces. We were still looking at the same Veronese surface, sometimes with precisely the same views, but we were led to totally different sorts of conjectures when we saw them in new ways. It was quite a surprising experience for a "pure" mathematician to see to what degree the most natural questions were being determined not by abstract criteria intrinsic to the mathematical subject but rather by the very practical matters of what it was that happened to be easiest to see at the time. "Because the light was better," we were following new leads and exploring directions we had not even considered at earlier stages of the process.

Methods which work well for our research inevitably have an impact on the way we teach, and the computer is destined to make a huge difference in which we will present the geometric aspects of our subjects. My own experience in this area has developed over the past fifteen years as equipment and student background have changed. For a long time my undergraduate courses in differential geometry have profited from slides and films and videotapes, but I wasn't prepared for the tremendous advance that came when the students were able to work interactively

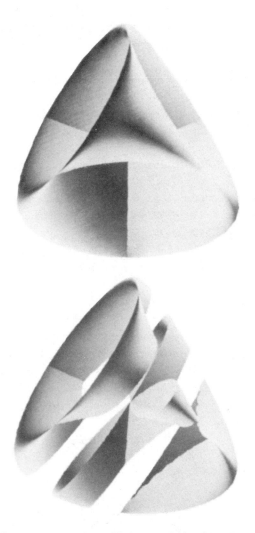

Figure 4. The Veronese surface, filled-in mode (top), and filled-in mode with slices (bottom).

with computers in a specially outfitted laboratory classroom using software generated by students Richard Hawkes, Timothy Kay and Edward Grove. Geometry truly

became an observational science. In the best circumstances, conjectures and theorems flow naturally from the phenomena which have been observed and discussed by the students themselves. A similar experiment in complex analysis, using courseware generated by student assistants Herbert Matteson and Richard Schwartz, has produced a very different attitude on the part of the participants than that which the author recalls from complex analysis courses taught with a minimum of visual information. The next breakthrough will occur when such interactive programs become available for small personal machines which are destined to become as available as electric typewriters for a previous generation. Our courses can build geometric intuition on an entirely different scale, and topics that had to be shortchanged for lack of adequate motivating examples can once again play an important role. Just as research follows different directions, so will the teaching change as the light improves.

So in my mathematical lifetime, all my old benchmarks that distinguished the pure from the applied have disappeared, sometimes in quite dramatic fashion. More and more mathematicians of all sorts find themselves inspired by phenomena that arise from computer graphics. They visit laboratories, collaborating with computer scientists and with students who take for granted subject matter which simply did not exist a generation ago. Under the circumstances, it often happens that some of the "pure" mathematical projects end up having rather direct relationships with applied problems. Occasionally the projects that arise in such situations attract funding, and possibly even consultations, a pleasant prospect for the future of our profession.

My final illustration for this presentation is the most recent film which has been made at Brown University in collaboration with Hüseyin Koçak, on the faculty of Applied Mathematics, and two Masters candidates in Computer Science, David Laidlaw and David Margolis. The film depicts stereographic projections of a family of toroidal surfaces in the 3-sphere, all of which have the two-piece property. In one sense it is the remake of the very first film which Charles Strauss and I made over fifteen years ago and it is appropriate for this occasion for another reason since one of its creators, David Margolis, received his first inspiration in mathematics as an undergraduate in Rochester as a student of the man we are honoring in the symposium, Gail Young.

Sequence from *The Hypersphere*: Rotating Hopf Tori

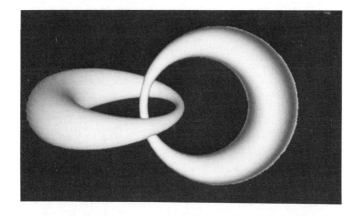

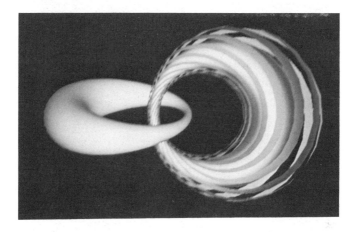

The Hypersphere

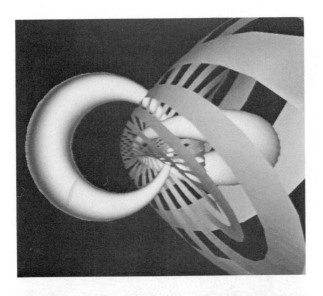

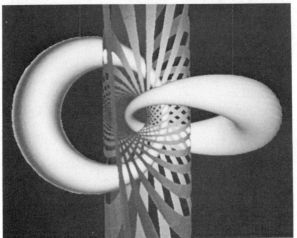

The Hypersphere

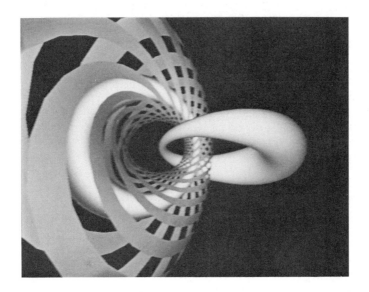

References.

[1] T. Banchoff, "Tightly embedded 2-dimensional polyhedral manifolds," *Amer. J. Math.*, **87**(1965), 462–472.

[2] T. Banchoff, "The spherical two piece property and tight surfaces in spheres," *J. Differential Geometry*, **4**(1970), 193–205.

[3] T. Banchoff, "The two-piece property and tight n-manifolds-with-boundary in E^3," *Trans. Amer. Math. Soc.*, **161**(1971), 259–267.

[4] T. Banchoff, "Non-rigidity theorems for tight polyhedral tori," *Archiv der Mathematik*, **21**(1970), 416–423.

[5] T. Banchoff and C. Strauss, "Real time computer graphics techniques," *Proceedings of Symposia in Applied Mathematics*, Vol. 20 (The Influence of Computing on Mathematical Research and Education) American Math. Soc., 1974, 105–111.

[6] T. Banchoff, "Computer animation and the geometry of surfaces in 3- and 4-space," *Proceedings of the International Congress of Mathematicians*, Helsinki, 1978 (invited 45 minute address), 1005–1013.

[7] T. Banchoff, T. Gaffney and C. McCrory, *Cusps of Gauss Mappings*, Pitman Advanced Publishing Program 55, London, 1982, 1–88.

[8] T. Banchoff and W. Kühnel, "The nine-vertex complex projective plane," *Mathematical Intelligencer*, **5**(1983), 11–22.

[9] T. Banchoff, S. Feiner and D. Salesin, "DIAL: a diagrammatic animation language," *IEEE Computer Graphics and Applications*, **2**(7)(1982), 43–54.

[10] T. Banchoff, "Differential geometry and computer graphics," *Perspectives in*

Mathematics, Anniversary of Oberwolfach 1984, Birkhäuser Verlag, Basel, 43–60.

[11] T. Banchoff, "Visualizing two-dimensional phenomena in four-dimensional spaces," in *Statistical Image Processing*, Marcel Dekker, to appear.

[12] T. Banchoff, H. Koçak, F. Bisshopp and D. Laidlaw, "Topology and mechanics with computer graphics: linear Hamiltonian systems in four dimensions," *Advances in Applied Mathematics*, to appear.

Thomas Banchoff's background and interests are detailed in this article. Briefly stated, he studied at Notre Dame and received a Ph.D. from Berkeley in 1964 under S. S. Chern. He was a Benjamin Peirce Instructor at Harvard for two years and spent a Fulbright year with Nicolaas Kuiper in Amsterdam before joining the faculty at Brown in 1967 where he has been ever since except for a sabbatical year at Berkeley and U.C.L.A., and another at the I.H.E.S. He has been involved in using computer graphics in geometry research and teaching since 1968.

MODELLING AND ALGORITHMIC ISSUES
IN INTELLIGENT CONTROL

Christopher J. Byrnes

Department of Electrical and Computer Engineering
Department of Mathematics
Arizona State University
Tempe, Arizona 85287

1. Introduction.

Despite ubiquitous success in the implementation of classical automatic control, there are pressing needs on many technological fronts for the design of more advanced, high performance, real-time command generators. For example, the needs for a significant increase in the accuracy, speed and versatility of robotic manipulators and of high performance weapon pointing systems have led to a reexamination of the implementation of classical (e.g., PD) controllers for DC actuators and an exploration and evaluation of the use of new and more sophisticated control schemes (see e.g. [1]–[8]). Aside from specific needs to meet more demanding performance requirements, more versatile command generators are now required to fully realize the benefits of the new design options which have been made possible by recent hardware innovations, ranging from devices such as microprocessors to DC motors. Indeed, recent advances in DC motor technology have made the implementation of direct drive actuators for robot arms attractive and feasible: the first of two new kinds of DC motors, based on rare earth cobalt magnets, has already been used in the Carnegie-Mellon direct-drive arm in 1981 and in the MIT direct drive arm (see [9]) in 1982, while a second kind of DC motor is currently being used in the construction of a four degree of freedom robot arm at the ASU Robotics Laboratory. The advent of direct drive actuators will allow robot arm motion which is an order of magnitude faster than previous conventional arms, with end effector speeds of up to 30 feet per second and accelerations of up to 5 to 7 G's, bringing robot motion control out of the quasi static domain and into a more complex dynamic domain. For example, in path planning for a robot arm using a more advanced actuator, the control design will need to minimize the effect

15

of structural resonances which will be excited as the servo bandwidth approaches the order of the structural natural frequencies. Similar performance improvements can, of course, be anticipated for other systems which are actuated by DC motors, requiring however replacing conventional PD controllers by command generators capable of real-time operation over wider bandwidths, or in nonlinear dynamical regimes, while meeting sophisticated objectives which may themselves require some on-line learning. Such performance requirements and new design opportunities, for these and for other kinds of systems (e.g. helicopter autopilots, intelligent control of silicon crystal growth, weapon pointing systems, etc.), underscore the fundamental importance of developing the foundations for methods for the design, analysis and implementation of intelligent controllers.

Since it is to these problems and opportunities, both in a broad and in a specific way, that research in intelligent control will be addressed, it is of increasing importance to examine the fundamental mathematical and system theoretic problems and principles which will underlie progress in intelligent control. Among the broader, more exploratory goals, is the development of intelligent command generators truly capable of operating in real-time to meet versatile needs in a variation of environments. As in some earlier descriptions of intelligent controllers (see, e.g., [8]–[11]), these control structures will likely ultimately possess a hierarchical structure (Section 2.1) with simple feedback servomechanisms at the lowest level and intelligent search at the higher, or supervisory levels. Central to the design of the supervisory level presented here is a consistency or "realization" check, based on the notion of "a priori information"; viz., precisely what system-theoretic information is required to achieve a control objective in real time and how can this information be extracted or learned in real time from the process description to achieve this objective. Our control structures will also be distinguished by the representation and design of intelligent controllers as knowledge-directed search procedures, with a search space prescribed by the supervisory level. Thus among our objectives is the development of knowledge-directed real time search procedures for the implementation of a structured, simple class of feedback control laws based on the development of a set of sufficiently powerful heuristics for linear and ultimately for nonlinear (cf. [16]–[18]) control. A specific, interesting but longer range, application would be the successful implementation of an intelligent controller for real-time path control of robotic manipulators in situations, such as conveyor coordination, where the path itself changes or needs to be learned during the arm motion.

In this paper, we outline a research program whose goal is the implementation of intelligent command generation in real-time. This program requires the solution of a variety of algorithmic and modelling problems which we feel are fundamental to the emerging field of intelligent control. First, the basic question of realizability (cf. Section 2) has so far remained unformulated and unanswered. This question deals with the very existence of a command generator capable of achieving the desired control objective given the available a priori information about the process to be controlled. This theory of "realization with constraints" has not even been formulated when the system (to be controlled) is known. For example, such a theory should be capable of asserting, on general principles, the existence and linearity of Luenburger observers, which we currently know exist only by virtue of an ad hoc construction. In the controller structure we propose here (Section 4), and illustrate in a case study (Section 5) adapted from [23], a consistency or realization check is performed at the supervisory level using system theoretic heuristics. For realizability by smooth nonlinear controllers, methods from nonlinear dynamics (see Section 6) provide a starting point for the development of such a realization theory, verifying the heuristic principle stated in Section 3, and used in the case study controller design in Section 5. Based on such calculations the general realization problem and the related "representation problem" (representing the controller as a knowledge-directed search, cf. Section 2) would seem to involve an intriguing but difficult blend of methods from mathematical system theory, nonlinear dynamics and artificial intelligence. The formulation and development of this basic theory is, in fact, one of our primary research programs, about which we expect to have more to say in the near future.

Another fundamental problem in intelligent control is the explicit design of search methods synthesizing the real-time aspects of dynamical systems with the back-tracking, interactive capabilities of intelligent problem-solving. Despite some rather intriguing examples and simulations, say for the prototypical problem of stabilization of a linear control system about a set point, there remains the need to develop intelligent command generators capable of actual real-time performance. In this paper, we describe some existing intelligent controllers which exhibit real-time learning in the form of a search (see [24]) implemented as an "ergodic" dynamical system on a set of classical control laws, similar in some ways to a Monte Carlo approach. Another of the basic programs outlined here, for set-point control and for more involved control schemes, is to maintain this dynamic learning characteristic while improving the real-time performance of these knowledge-directed searches by incorporating search methods, where appropriate, from artificial and natural intelligence in systematic yet a rigorous fashion.

As an example, one intelligent counterpart of the PID controllers used so commonly in the control of "quasi-static" robot arms consists of a knowledge-directed, dynamical search through the PID parameter space using the Ziegler-Nichols rules in conjunction with control-theoretic heuristics such as autotuning [7]–[8], or with a heuristic search method, such as simulated annealing. Another intelligent controller, for system stabilization, which exhibits these learning characteristics together with backtracking to stabilizing feedback laws of minimal norm has just been developed ([25]) and will be reported elsewhere.

Finally, let us remark that, in order to develop intelligent command generators capable of real-time higher performance operation in a variation of enviroments, it will become necessary to be able to use analysis and design principles which apply to, at least to some broad class of, nonlinear systems as well as linear systems. In [16]–[18] a research program is initiated whose goal is the development of heuristics for nonlinear control, similar in scope and spirit to classical control, to be used in the analysis and design of nonlinear feedback control systems (see also [19]–[22] for earlier work by which this program was partially inspired). In particular, this includes developing nonlinear generalizations of some of the concepts familiar from frequency domain control theory and using these, in much the same way as classical control methods, to design and analyze nonlinear versions of PD control, lead-lag compensation, etc., to "shape the response" of nonlinear systems. The utility of such methods in intelligent control is based in part on having provided a set of sufficiently powerful system-theoretic heuristics to permit implementation of real-time searches and in part on the development of a class of simply structured control laws capable, for example, of stabilizing a nonlinear system given only a crude knowledge of the actual system parameters. In practice these parameters typically consist of literally thousands of transcendental functions, rendering the on-line parameter estimation of the system coefficients an extremely unattractive alternative, from the point of view of both rigorous analysis and cost effectiveness. We also believe this is useful because the simpler the design philosophy and the simpler the controller structure, the more likely it is that the controller can be "molded" to fit a particular application. This formulation of Occum's razor will also be one of the guiding principles in our discussion of intelligent controllers for linear systems.

I would like to thank K.-J. Åström, C.-Y. Kuo, B. Mårtensson and J. C. Willems for interesting and very useful discussions, suggestions and collaborations on the program in intelligent control which is outlined in this paper. I would also like to thank Dick Ewing, Ken Gross and Clyde Martin for giving me the opportunity to present these ideas to an interesting audience gathered to congratulate Gail Young for his lasting contributions to American mathematics.

2. The Problem Structure.

In order to fix the ideas, we begin with a fairly broad description of some of the kinds of control problems where intelligent control is required, emphasizing the points in the command generation process where in fact some kind of learning or intelligence is necessary.

We delineate the problem structure as follows:

a. *The Process to be Controlled.* The first data consists in a perhaps rough description of the system to be controlled. That is, we are to control an input-output process where, in lieu of an explicit parametric description of σ, we instead represent

Figure 2.1. Depicting the I/O process to be controlled.

our knowledge of the input-output model by membership, $\sigma \in \Sigma$, in some fixed class of (linear or nonlinear, finite or infinite dimensional) systems, perhaps modelling a nominal plant together with structured or unstructured perturbations.

For example, suppose σ consists of the rigid-body model

$$M(q)\ddot{q} + B(q,\dot{q})\dot{q} + K(q) = T \qquad (2.1)$$

for the dynamics of a robot arm. Thus, the inertia terms $M(q)$, the velocity terms $B(q,\dot{q})$ containing e.g. Coriolis and centripetal forces, and the gravitational terms $K(q)$ contain literally thousands of nonlinear terms, not all of which can be assumed to be known precisely and some of which vary due to environmental changes and to effects of ignored (high frequency) actuator dynamics. In this case, one choice for Σ would consist of the class of models (2.1) with some problem-specific qualitative or quantitative constraints on M, B, K.

b. The Control Objective. Typical examples where the control objective is known a priori are set-point control for the output (such as the position of a manipulator end effector)

$$e(t) = y(t) - y_d \to 0, \quad t \to \infty,$$

or path following (path-planning), where the path error should decay asymptotically

$$e_r(t) = y(t) - y_r(t) \to 0, \quad t \to \infty,$$

together with its derivatives ($\dot{e}$, $\ddot{e}$, etc.). Thus, while the control objective must be achieved in real-time, despite a perhaps significant lack of precision in the model description, the objective is described in terms of quantities (i.e., y_d, $y_r(t)$) known in advance. A much harder problem arises (e.g. in conveyor belt coordination) when, in addition, the desired trajectory must also be learned in real-time. We expect the mathematics of this harder problem to consist, e.g. in the on-line identification or "learning" of quantitative or qualitative information concerning a differential equation and initial data

$$f(\ddot{q}_r, \dot{q}_r, q_r) = 0, \quad q_r(0), \quad \dot{q}_r(0)$$

defining the trajectory.

c. Realization of an Intelligent Command Generator Achieving the Desired Objective. Deferring questions of rigorous analysis, performance and implementation to later sections, we refer here to a basic, fundamental problem which has however remained unformulated in the emerging field of intelligent control. Namely, despite some rather intriguing and promising advances in the design and modelling of intelligent controllers, rather little attention has been paid to the very basic question of existence, or realizability, of such controllers once (a) and (b) have been specified:

Realization Problem for Intelligent Control: Does there in fact exist a command generator (modelled, e.g., as a finite or infinite dimensional, linear or nonlinear system) achieving the objective specified in (b) for the process models specified in (a).

In the controller structure we discuss in Section 4, the Realization Problem will be the focus of a supervisory level, where solvability would lead to the designation of a "search set" of candidate, simply, structured feedback laws thereby solving, in the language of AI, the "representation problem." Production rules will then take the form of a knowledge-directed search, prompted by closed-loop performance, on a set of simply structured command generators.

3. Examples.

We now give some examples, illustrating (a) and (b) described in the definition of the problem structure (Section 2) as well as what we would mean by an intelligent controller.

Example 3.1 (First-Order Linear Systems with Unknown Dynamics). In this simple example, we take the class of processes to be controlled (a) for Σ the class of first-order systems with positive gain.

$$\Sigma = \{\dot{y} = ay + bu, \quad a, b \text{ unknown}, t > 0\} \tag{3.1}$$

For the control objective (b) we take stabilization about the set-point $y_d = 0$. Thus, we seek a command-generator which would stabilize the closed-loop system for any

Figure 3.1. Depicting the command generator to be realized.

$\sigma \in \Sigma$; i.e., for any $a, b, b > 0$. Classical automatic control would suggest the feedback law

$$u = -ky, \quad k > a/b \qquad (3.2)$$

where a/b is assumed known, while standard adaptive control would suggest estimating the parameters a, b, or a/b. On the other hand, system-theoretic heuristics would suggest modifying (3.1) to obtain, for example, a nonlinear version of proportional error feedback:

$$u = -ky, \quad \dot{k} = y^2 \qquad (3.3)$$

This scheme, in fact, does work. Specifically

Proposition 3.1. *The closed-loop system (3.1)–(3.3)*

$$\dot{y} = ay - bky \qquad (3.4a)$$
$$\dot{k} = y^2 \qquad (3.4b)$$

satisfies

$$\big(k_0, y_0\big), \quad k_t \to k_\infty, \ y_t \to 0 \quad \text{as } t \to \infty$$

Proof. Differentiation along the trajectories of (3.4) shows that,

$$H(k, y) = y^2 - 2ak + bk^2$$

is a conserved quantity; i.e. $H(k,y)$ is constant along trajectories which therefore evolve along the (semi-)ellipses defined by

$$H(k, y) = \text{const.} \qquad (3.5)$$

$$\text{Q.E.D.}$$

The trajectories are depicted in Figure 3.2, below. We remark that, although one may consider the stabilization of (3.1) as a problem in adaptive control, the controller (3.3) is not based on a standard parameter indentification algorithm but illustrates instead a new and rather "orthogonal" approach ([19]–[20]) to this basic problem: as is depicted in Figure 3.2, the controller (3.3) "learns" enough about (3.1) in real-time to adjust itself to a parameter value k_∞, $k_\infty > a/b$, which a control engineer would have chosen on the basis of classical automatic control, cf. (3.2). In this way, (3.3) can be interpreted as a knowledge-directed search through a space, viz. R^+, of classical compensators. In [20]–[??], it is shown quite forcefully that the general stabilization

22

problem can be solved using such a search.

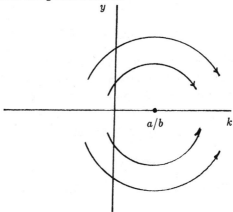

Figure 3.2. The phase portrait of (3.1)–(3.3).

Example 3.2 (Intelligent Critical Damping of a DC Motor). We consider a linear DC motor actuating a rigid planar joint, with input torque T and output the deflection θ of the joint from the x-axis. Thus, a transfer function model for σ is given by

$$g(s) = \frac{1}{Js^2 + Fs} \tag{3.6}$$

while in state space, we have

$$\begin{bmatrix} \dot{\theta} \\ \ddot{\theta} \end{bmatrix} = \begin{bmatrix} 0 & 1 \\ 0 & -F/J \end{bmatrix} \begin{bmatrix} \theta \\ \dot{\theta} \end{bmatrix} + \begin{bmatrix} 0 \\ 1/J \end{bmatrix} T \tag{3.7a}$$

$$\theta = [\,1, 0\,] \begin{bmatrix} \theta \\ \dot{\theta} \end{bmatrix} \tag{3.7b}$$

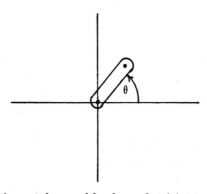

Figure 3.3. Depicting a 1-degree of freedom robot joint, actuated by a linear DC motor.

Motivated by the control of manipulators with multiple joints, where the forces on a single joint depend on the configuration and nonlinear coupling of the other joints, we take for (a) the class of systems

$$\Sigma = \left\{ \frac{1}{Js^2 + Fs} : J, H > 0 \text{ but unknown} \right\}$$

Taking, for the control objective (b), set-point control about $\theta_d = 0$, the nonlinear proportional error feedback law (3.3)

$$T = -K\theta, \quad \dot{K} = \theta^2 \tag{3.8}$$

still stabilizes the closed-loop system, in the sense that

$$V K_0, \ \theta_0, \ \dot{\theta}_0, \ (\theta_t, \dot{\theta}_t) \to 0, \ K_t \to K_\infty \quad \text{as } t \to \infty$$

While a rigorous convergence analysis of this scheme is possible, it is in fact far more involved than the analysis in Example 3.1. More generally, one can show:

Proposition 3.2. *The class Σ of linear systems defined via the transfer function description:*

$$\Sigma = \left\{ \frac{1}{s^2 + a_1 s + a_2} : a_1 > 0 \right\}$$

is stabilized by fixed control strategy (3.3). That is, the closed-loop system

$$\dot{x}_1 = x_2$$
$$\dot{x}_2 = -(a_2 + k)x_1 - a_1 x_2$$
$$\dot{k} = x_1^2$$

satisfies, for all initial data (x_0, k_0), $x_t \to 0$, $k_t \to k_\infty$ as $t \to \infty$.

Intuitively, treating the tuned parameter k_t as a time-independent constant, the root-locus plot for the closed loop system takes the form

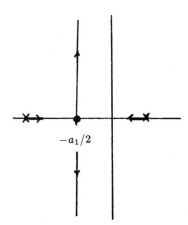

$-a_1/2$

so that for $k \gg 0$, the closed-loop eigenvalues lie in $\bar{}$. Heuristically, then the feedback law (3.3) will stabilize the closed-loop system provided k_t remains bounded.

Proof. First note that, if k_t is bounded, then $k_t \to k_\infty$ as $t \to \infty$, for some k_∞ depending perhaps on the initial data. Therefore, if k_t is bounded, $x_1 \in L^2(0, \infty)$. In particular, the system

$$\dot{x} = -(a_2 + k)x_1 - a_1 x_2$$

consists of a stable, autonomous linear system being driven by an L^2 forcing function and therefore x_2 belongs to $L^2(0, \infty)$. But then, $\dot{x}_1, \dot{x}_2 \in L^2(0, \infty)$ and we must have

$$(x_1)_t, \; (x_2)_t \to 0 \quad \text{as } t \to \infty.$$

Lemma. k_t *remains bounded.*

Sketch of Proof. (due to W. Cebuhar) Suppose not. Consider the Lyapunov function

$$N(x_1, x_2, k) = \left(\frac{a_1}{2} + \frac{a_2 + k}{a_1}\right)x_1^2 = x_1 x_2 + \frac{1}{a_1}x_2^2$$

which is nonnegative for $k(t) \geq |a_2|$ and note that, along the closed-loop trajectories

$$\dot{v}(x_1, x_2, k) = -(a_2 + k)x_1^2 - x_2^2 + \frac{x_1^4}{a_1}$$

Choosing t_0 such that

$$k(t) \geq \frac{a_1^2}{4} - a_2 \quad \text{for } t \geq t_0$$

and the time-varying change of coordinates

$$z = \exp\left(\frac{a_1}{2}(t - t_0)\right)x_1$$

we have

$$\ddot{z} + p(t)z = 0$$

where $p(t)$ is divergent monotone nondecreasing. This implies, however, that z_t and hence $(x_1)_t$ is bounded. Integration of $\dot{V}$ now shows that consequently, k_t must be bounded, giving a contradiction. Q.E.D.

As a second example of (b), much more salient to end-effector control, we consider the problem of designing an intelligent controller which critically damps (3.6), having to learn the unspecified inertia and friction terms. Thus, we would seek a knowledge-directed search—perhaps modelled by a nonlinear process

$$T = -k\theta, \quad \dot{k} = f(k, \theta)$$

such that when $k_t \to k_\infty$, "the terminal closed-loop dynamics"

$$\frac{1}{Js^2 + Fs + K_\infty}$$

is critically damped, i.e.

$$F^2 - 4K_\infty J = 0$$

despite the lack of a priori knowledge of F, J. In this comparatively simple setting, even the realization, or representation problem seems interesting.

Example 3.3 (Path Planning for a Nonlinear Manipulator). We first consider the problem (b) of path planning for a robotic manipulator taking for (a) the model (2.1), where the reference path

$$q_r(t), \ \dot{q}_r(t), \ \ddot{q}_r(t) \tag{3.9}$$

is known in advance. In this case, an algorithm for path-following is given by the standard method known variously as "inverse dynamics," "feedback linearization," or "computed torque," see e.g. [3], [28], [30]. Explicitly, assuming as one may that $M(q)$ is invertible, (2.1) can be expressed as:

$$\dot{q}_1 = q_2 \tag{2.1a}'$$

$$\dot{q}_2 = -M^{-1}(q_1)\big(B(q_1, q_2)q_2 + K(q_1) - T\big) \tag{2.1b}'$$

If the torque T is computed as (or generated via)

$$T = -K(q_1) - B(q_1, q_2)q_2 + M(-q_2 - q_1 + \ddot{q}_r + \dot{q}_r + q_r) \tag{3.10}$$

then the closed-loop system takes the form

$$\dot{q}_1 = q_2$$

$$\dot{q}_2 + q_2 + q_1 = \ddot{q}_r + \dot{q}_r + q_r$$

and hence

$$q(t) - q_r(t) \to 0 \quad \text{as } t \to \infty.$$

While incorporating the nonlinear terms in (3.10) seemingly allows one to bypass the parameter variation questions (cf. Example 3.2) in linear models such as (3.6), feedback laws such as (3.10) explicitly rely on exact cancellation of the literally thousands of nonlinear contributions to the terms $M(q)$, $B(q, \dot{q})$ and $K(q)$, not all of which can be considered to be known with the required precision. Moreover, since these terms are polynomials in transcendental functions and therefore contain Taylor coefficients of arbitrarily high order, real-time control would seem to preclude the use (for nonlinear systems) of a parameter identification subroutine. Indeed, controllers which rely on intensive on-line numerical processing would seem less likely to offer a cost-effective controller design for the high performance, wide bandwidth systems projected for the next generation of process control and military systems.

Thus, in contrast to the problem described in Examples 3.1–3.2, it cannot be expected that a controller for systems such as (2.1) would be designed on the basis of learning all of these thousands of complex nonlinearities in real-time. Instead, we propose the development of a class of intelligent controllers using a knowledge-directed search of a class (defined on a problem-specific basis) of simply structured controllers, employing a sufficiently powerful set of system-theoretic heuristics valid for both linear and nonlinear control systems.

4. The Structure of the Command Generator.

We now describe the development of a class of control structures which would employ knowledge-directed real-time search procedures for the implementation of a set of a simple class of feedback control laws. Typical of applications of AI to real-time control (see, e.g., [8]–[11], [25]), these control structures will ultimately possess a hierarchical structure with simple feedback servomechanisms at the lowest level and intelligent search at the higher, or supervisory, levels. Our control structures will, however, be distinguished by the representation and implementation of intelligent controllers as knowledge-directed search procedures. The feasibility of search based controllers has been demonstrated quite convincingly in the "iterative control" approach to the development of learning control systems for robotic manipulators ([4]–[5], [27]–[28]) and in the prototypical set-point control problem for plants with parametric uncertainties in [23]–[24]. In [23]–[24] the controller design is based on the notion of "a priori information": viz., precisely what system-theoretic information is required to achieve a control objective in real time and how can this information be used or learned in real time to achieve this objective. Quite orthogonal to conventional parameter adaptive controllers, these algorithms exhibit "learning" in a manner which is also different from AI searches using production rules: the learning process, or search, is realized as a "quasi-ergodic" dynamical system defined on an appropriate space of compensators, with ω-limits which are compensators a control engineer would have designed were the system parameters known in advance.

It is this aspect of the algorithms in [23]–[24] which we want to retain in our design of intelligent command generators. We feel that such dynamical system implementations of heuristic searches is an extremely promising research area, which is ultimately capable of forming the basis for real time implementation of intelligent controllers.

To fix the ideas, we will now suppose, as in Section 2, the following problem data is given: (a) a class Σ of m-input, p-output systems, and (b) an objective. Our intelligent command generator would roughly proceed along the following hierarchical steps:

I. A consistency or realization check at the supervisory level; checking perhaps on a heuristic basis if the information defining Σ is sufficient for the existence of a real-time command generator meeting the control objective;

II. The representation problem; the specification, based on the consistency check in (I), of a "search set" of simply structured command generators meeting this control objective;

III. The design and the closed-loop implementation of a real-time, knowledge-directed search on this set, based on a sufficiently powerful set of system-theoretic heuristics, driven by closed-loop performance of the candidate command generators.

In order to further illustrate the ideas, we fix the control objective, choosing stabilization about a set-point, and restricting (cf., however [16]–[18]) attention to the control of linear systems. The consistency check thus amounts to the basic question:

(Q1) What a priori information is needed to (universally) stabilize a given class Σ of linear systems?

In Section 6 we will sketch a rigorous verification (involving notions from nonlinear dynamics, e.g. center manifold theory) of the following heuristic principle related to question, (Q1):

(H1) A fixed controller exists stabilizing each $\sigma \in \Sigma$, provided from the description of Σ one can logically deduce the existence of a natural number q such that: each fixed $\sigma \in \Sigma$ can be stabilized by some linear compensator of dimension q.

Thus, using predicate calculus and a library of system-theoretic facts or rules, production rules for a consistency check would be of the form:

description of Σ $\quad - - \longrightarrow \quad$ bound on the order of stabilizing compensators

(antecedent) $\quad - - \longrightarrow \quad$ (consequent)

For example if

$$\Sigma = \{\sigma : \deg \sigma \leq n\}$$

then we have

each system σ has degree $\leq n$ $\quad - - \longrightarrow \quad$ $q = n - 1$

(antecedent) $\quad - - \longrightarrow \quad$ (consequent)

because of the rule, "any linear system of degree n can be stabilized by a linear compensator of order $\leq n - 1$."

This illustrates the use of heuristics in (I). For the control objective (II) we have fixed above, we have the L^2-stability heuristic

(H2) If a linear system is stable, then for all initial data x_0 and all $T \geq 0$

$$\int_0^T \|x_t\|^2 \, dt \leq c\|x_0\|^2$$

where c is a constant depending only on the system.

In Section 5 we will illustrate the implementation of (H1)–(H2) for the intelligent set-point control of a particular class of system. A rigorous justification for the implementation of these heuristics is sketched in Section 6.

There are, of course, other choices of heuristics for set-point control. For example, it is classical to obtain satisfactory set-point control for low-order systems by using PD or PID controllers, where the parameters K_e, K_I, and K_D, of the controller are calculated, for example, from the Ziegler-Nichols rules, using an approximate determination of the intersection of the Nyquist curve with the real-axis. This heuristic has recently been used by Åström and Hägglund ([8]) as the basis of an intelligent control scheme, called "autotuning," which roughly "learns" this intersection point in real-time by driving the system with a relay, inducing the system to oscillate in a limit cycle.

Returning to a general discussion of the structure (I–III) of an intelligent set-point controller, we first note that an affirmative verification of the supervisory consistent check (I) using (H1) would lead in step (II) to the prescription, as a search space, of the manifold (see e.g. [23]–[24]) K_q of compensators of dimension equal to q. Further information concerning Σ might be used at this point to identify a "smaller dimensional" search space; i.e., some subset (perhaps a submanifold) of K_q.

Is it now possible to define (as is required in Step III) a real-time search on K_q leading to closed-loop stability? The answer is, perhaps surprisingly, in the affirmative. B. Mårtensson [24], [29] has constructed a search in K_q, modelled by a smooth ergodic flow with the property:

If every system in Σ can be stabilized by some linear compensator of order $\leq q$, implementing the ergodic search on K_q in closed-loop leads for each $\sigma \in \Sigma$, to internal stability.

Mårtensson's search is driven by a modification of (H2), viz. by a heuristic based on L^2-stability:

(H2)′ If a linear system is stable, for all square integrable inputs (e.g. exponentially decaying) u_t

$$\int_0^T \left(\|u_t\|^2 + \|y_t\|^2 \right) dt < \infty$$

Thus, this "ergodic" search is initialized at an arbitrary compensator $K_0(s)$ and learns, prompted by closed-loop performance, enough in real-time to adjust itself to a compensator $K_\infty(s)$ stabilizing the unknown plant. Simulations of this search unfortunately require an extremely vast amount of storage and computing time and, for these reasons, in its broadest generality implementing Mårtensson's algorithm might be compared ([29]) to implementing an effective computation on a Turing machine. However, as the case-study in Section 5 shows, by incorporating more precise problem-specific knowledge concerning Σ with a powerful set of system theoretic heuristics, it is indeed possible to greatly improve closed-loop performance.

5. A Case Study: An Intelligent Set-Point Controller.

We now illustrate the hierarchical control structure outlined in Section 2.3 in a case-study, the class of scalar linear control systems:

$$\Sigma = \{g(s) : g(s) \text{ minimum please, relative degree 1}\}$$

Recall, $g(s)$ is rational, $g(s) = p(s)/q(s)$, and the relative degree of $g(s)$ is defined by $\deg q(s) - \deg p(s)$.

In particular no assumption is made on the system dimension. In this case, the intelligent command generator (I)–(III) takes the form:

(I) The consistency check at the supervisory level.

At this level, assuming validity of the heuristic (H1), we could record an affirmative consistency check

each σ is minimum phase relative degree one	$- - \rightarrow$	each σ can be stabilized by a compensator of order $q = 0$
(antecedent)	$- - \rightarrow$	(consequent)

using the classical, root-locus based production rule, "every minimum phase system of relative degree r can be stabilized by a compensator of order $q \leq r - 1$."

(II) The representation problem.

The supervisory level check then prescribes the search space $K_0 = R$ of constant gain $(q = 0)$ direct output feedback laws.

(III) The design and implementation of a real-time knowledge-directed search.

Quite generally, at level (II) we have a prescription of K_q (or a subset thereof) as a search space. Choosing a countable dense grid $\{K_i\} \subset K_q$ and representing K_i as an operator on the output $y(t)$, we would, for example, employ the heuristic (H2) in some knowledge-directed search. One such search (suggested tome by Mårtensson) may be expressed in a fictitious PASCAL-like language, as:

> For $k := 1$ to infinity do
>> For $i := 1$ to k do
>>
>> **begin**
>>> $t_0 := t$
>>>
>>> $\|x(t_0)\|^2 := \|x\|^2$
>>>
>>> $K := K_i$
>>
>> **repeat**
>>> $u := k * K_i * y$
>>
>> **until**
>>
>> $$\int_{t_0}^{\infty} \|x(t)\|^2 \, dt > k * \|x(t_0)\|^2$$

For the specific description of Σ given above, we can modify (H2) by the stability criterion:

(H2)″ $\int_0^T y(t)^2 \, dt < c y_0^2$

since each $\sigma \in \Sigma$ is minimum phase. We now implement this controller in real-time as follows (see [23]). The integral in our program can then be replaced by the real-time quantity

$$\int_0^T y(t)^2 \, dt$$

which suggests using a sampled version of the real-time, dynamic search

$$\dot{k} = y^2 \qquad (5.1)$$

where the feedback law is then implemented by

$$u = \pm k y$$

depending on the sign of the system instantaneous gain. This then suggests implementing the feedback law, of "Nussbaum gain,"

$$u = s(k) k y \qquad (5.2)$$

where on the average $s(k)k$ looks both like $+k$ and like $-k$; i.e., the Cesaro mean of $s(k)k$ should satisfy

$$\limsup \frac{1}{k} \int_0^k s(T) T \, dT = +\infty, \qquad \liminf \frac{1}{k} \int_0^k s(T) T \, dT = -\infty$$

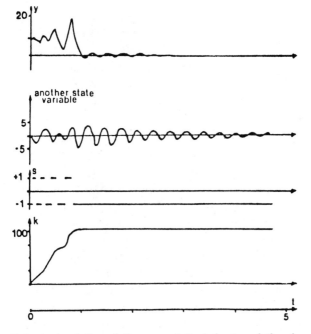

Figure 5.1. A simulation of the asymptotic behavior of the closed-loop system (5.1)–(5.4).

For example, we can take

$$s(k) = \begin{cases} +1 & n^2 \leq |\kappa| < (n+1)^2 \quad n = 0, 2, 4, \ldots \\ -1 & n^2 \leq |\kappa| < (n+1)^2 \quad n = 1, 3, 5, \ldots \end{cases} \tag{5.3}$$

Thus, implementing (5.1)–(5.2) gives a real-time representation of the knowledge directed search (III) as a dynamical system on K_0, illustrating the general construction guaranteed by (H1) and Mårtensson's algorithm. A simulation of the closed-loop behavior of this dynamic, knowledge-directed search is given in Figure 5.1 for the third order open-loop system

$$g(s) = \frac{.15s^2 + .3s + 6.2}{s^3 + 1.5s^2 + .36s - 22}$$

A rigorous convergence analysis for this controller is given in the next section.

6. Rigorous Analysis of the Heuristics.

In general, we anticipate the design, analysis and implementation of intelligent controllers described in Section 4 to involve two kinds of heuristics. Heuristic algorithms, of course, call for analytic justification not only for confidence in their use and a clear delineation of their scope, but also because a clear, analytic understanding of precisely what such algorithms are doing is fundamental for future progress in real-time intelligent control.

The first kind of heuristic, for example (H1), is employed at the supervisory level, for the most part to solve the representation problem, while checking whether the desired control objective can in fact be achieved in real-time given the a priori information about the process to be controlled. The second involves those system-theoretic heuristics, such as (H2), used to realize and to implement a real-time search. We now illustrate the kinds of mathematics which can be used to justify (H1)–(H2).

Even for set-point control, the representation problem and the related consistency check is intriguing and still very much open. Roughly speaking, (H1) asserts that any command generator (Figure 2.2) for the I/O process (Figure 2.1) can in fact be realized, or interpreted, as a search on a space K_q of compensators. The motivation for (H1) is the hypothesis, assuming a sufficiently high sampling rate, that the command generator can be realized as a sampled, continuous time but perhaps nonlinear, control system

$$\dot{z} = f(z, y)$$
$$u = g(z, y) \tag{6.1}$$

with y_t as input and u_t as output. We further assume (6.1) is smooth, i.e. that f, g are C^∞ functions of z in some R^q (or of some manifold M of dimension q). This is the case, for example, for the controller (5.1)–(5.2). With these working hypotheses, (H1) asserts that the q-dimensional nonlinear controller (6.1) can be interpreted as generating a dynamical search on the space K_q of linear compensators of dimension $\leq q$.

To illustrate the basic idea, we recall from Example 3.1 the phase portrait depicted in Figure 3.2. As depicted, the controller (3.3) learns enough about the system (3.1) in real-time to adjust itself to a limiting feedback law k_∞,

$$u = -k_\infty y$$

which a control engineer would have designed, were the system actually known. Thus the ω-limits of (3.1)–(3.3) give the ω-limits of a search on the space K_0 of constant gain feedback laws. As shown in Byrnes, Helmke and Morse [34], this asymptotic "learning" feature is in fact the case whenever the general controller (6.1) actually achieves set-point control; i.e. one can show the existence of a "good" ω-limit. If the ω-limits of the closed-loop system were isolated, this would follow from a Baire category argument and the Grobman-Hartman theorem from nonlinear dynamics. As examples and simulations indicate, this rarely seems to be the case. More precisely, smooth controllers (6.1) typically lead to closed loop systems with nontrivial center manifolds (see e.g. Example 3.1 and Section 5). This technical complication can be overcome using more sophisticated tools from nonlinear dynamics (e.g. center manifold theory, the reduction theorem, etc.) together with a Baire Theorem argument. From this, (H1) follows. We want however to reiterate our interest in the general representation question; i.e. when our hypotheses concerning smoothness of realizations for setpoint controllers do not hold. We expect this general case to involve a difficult but fundamental blend of discrete and continuous nonlinear mathematics in a new and important way.

A rigorous convergence analysis of searches prompted by heuristics such as (H2) typically involves the following dichotomy: while many classical system theoretic heuristics are frequency domain, the implementation is largely state-space. For this reason, the analysis of algorithms such as (5.1)–(5.4) reposes on a state-space generalization of frequency domain methods, such as root-locus theory, etc.

Since our development and analysis of heuristics for nonlinear control involves similar tools, we now illustrate some of these arguments rather explicitly. For σ as in (5.1)–(5.4) we can choose (following [19]) a minimal realization (A, b, c) so that the closed-loop system equations take the form

$$\dot{z} = A_{11}z + A_{12}y$$
$$\dot{y} = A_{21}z + A_{22}y + cbu, \qquad u = s(k)ky$$
$$\dot{k} = y^2$$

Here $(z^t, y^t)^t$ is a partition of the state variable $x \in R^n$ compatible with the partition

$$R^n = \ker c + \mathrm{sp}\{b\}$$

which is valid since the system has relative degree 1, i.e. since $cb \neq 0$. By a Laplace transform argument, we see

$$\mathrm{spec}(A_{11}) = \mathrm{zeroes}\big(g(s)\big) \tag{2.5.2}$$

and therefore A_{11} is a stable matrix. For simplicity, assume $cb > 0$ so that we may take $s(k) = -1$. We wish to show

(i) $z_t, y_t \to 0$ as $t \to 0$

(ii) $k_t \to k_\infty$

First, we note that (ii) implies (i). Explicitly, if (ii) holds, $y_t \in L^2(0, \infty)$ and therefore $z_t \in L^2(0, \infty)$ by L^2-stability of the auxiliary control system

$$\dot{z} = A_{11}z + A_{12}y$$

Therefore, by linearity, $\dot{z}_t, \dot{y}_t \in L^2(0, \infty)$ and hence

$$z_t, \ y_t \to 0$$

since the inner products

$$\langle z_t, \dot{z}_t \rangle, \quad \langle y_t, \dot{y}_t \rangle$$

are finite.

It remains, then, to prove (i). Assume $k_t \uparrow \infty$ and consider the time-varying version of a singularly perturbed system

$$\dot{z}_t = A_{11}z_t + A_{12}y_t$$
$$\dot{y}_t = A_{21}z_t + A_{22}y_t - k(t)cby_t.$$

Since $cb > 0$ and $\text{spec}(A_{11}) \subset C^-$, a combination of an L^2-stability and a Lyapunov argument shows that all solutions must decay exponentially. Therefore, $y_t \in L^2(0, \infty)$ and therefore $|k_t| \leq M$, contradicting $k_t \uparrow \infty$. Hence, $k_t \to k_\infty$ with k_∞ finite.

This curious combination of singular-perturbation methods and nonlinear dynamics extends to other linear and nonlinear control problems and forms part of the basis of our approach to the derivation of heuristics for nonlinear control.

Acknowledgment. This research was supported in part by AFOSR under grant 85-0224 and by NSF under grant NSF8513099.

References.

[1] P. T. Yip, "Microprocessor implementation and evaluation of a self-tuning control system," *Proc. Fifth Meeting of the U.S. Armament R/D Command*, Dover, New Jersey, 1983, 223–232.

[2] E. Freund, "A nonlinear control concept for computer controlled manipulators," *Proc. IFAC Symp. on Multivariable Technological Systems*, Fredericton, 1977.

[3] T. J. Tarn, A. K. Bejczy, A. Isidori and Y. Chen, "Nonlinear feedback in robot arm control," *Proc. of 23rd IEEE Conf. on Dec. and Control*, Las Vegas, 1984, 736–751.

[4] S. Arimoto, S. Kawamura and F. Miyazaki, "Bettering operation of dynamic systems by learning: a new control theory for servomechanism or mechatronics systems," *Proc. 23rd CDC*, 1984, 1064–1069.

[5] T. Mita and E. Kato, "Iterative control and its applications to motion control of robot arm – a direct approach to servo problems," *Proc. of 24th Conf. Dec. and Control*, Ft. Lauderdale, 1985, 1393–1398.

[6] M. Le Borgne, J. M. Ibarra and B. Espian, "Adaptive control of high velocity manipulators," *Proc. of the 11th Intl. Symp. on Industrial Robots*, Tokyo, October 1981, 227–236.

[7] K.-J. Åström, "Ziegler-Nichols auto-tuners," *Technical Report*, Lund Institute of Technology, May 1982.

[8] K.-J. Åström and T. Hägglund, "Automatic tuning of simple regulators with specifications on phase and amplitude margins," *Automatica*, **20**(1984), 645–651.

[9] Haruhiko Asada, "M.I.T. Direct-Drive Arm Project," *Robots 8 Conference Proceedings*, 16-10–16-21.

[10] G. Blankenship, J.-P. Quadrat et al., "An expert system for control and signal processing with automatic FORTRAN code generation," *Proc. of 23rd Conf. Dec. and Control*, Las Vegas, 1984, 716–728.

[11] G. Blankenship, J.-P. Quadrat et al., "An expert system for stochastic control and nonlinear filtering," *Theory and Application of Nonlinear Systems* (C. I. Byrnes and A. Lindquist, eds.), North Holland, 1986.

[12] K.-J. Åström, "Auto-tuning, adaptation and expert control," *Technical Report*, Lund Institute of Technology, October 1985.

[13] K.-J. Åström, J. Anton and K. E. Årzen, "Expert control," *Automatica*, **22**(1986), to appear.

[14] G. Saridis and J. H. Graham, "Linguistic decision schemata for intelligent robots," *Robotics Automation Lab. Report 6*, 1982.

[15] A. DeRochers, "Intelligent adaptive control strategies for hot steel rolling mills," *Proc. Yale Workshop on Applic. of Adaptive Syst. Theory*, 1981, 125–131.

[16] C. I. Byrnes and A. Isidori, "A frequency domain philosophy for nonlinear systems, with application to stabilization and to adaptive control," *Proc. of 23rd IEEE Conf. on Dec. and Control*, Las Vegas, 1984, 1569–1573.

[17] C. I. Byrnes and A. Isidori, "Asymptotic expansions, root loci and the global stability of nonlinear feedback systems," *Algebraic and Geometric Methods in Nonlinear Control Theory* (M. Fliess and M. Hazewinkel, eds.), D. Reidel Publishing Co., 1985.

[18] C. I. Byrnes and A. Isidori, "Global feedback stabilization of nonlinear minimum phase systems," *Proc. of 24th IEEE Conf. on Dec. and Control*, Ft. Lauderdale, 1985, 1031–1037.

[19] A. Isidori, A. J. Krener, C. Gori-Giorgi and S. Monaco, "Nonlinear decoupling via feedback, a differential geometric approach," *IEEE Trans. Aut. Control*, AC-**21**(1981), 331–345.

[20] R. Hirschorn, "(A, B)-invariant distributions and disturbance decoupling of nonlinear systems," *SIAM J. Control and Opt.*, **19**(1981), 1–19.

[21] R. Hirschorn, "Invertibility of nonlinear control systems," *SIAM J. Control and Opt.*, **17**(1979), 289–297.

[22] E. Freund, "Direct design methods for the control of industrial robots," *Computers in Mech. Eng.* (1983), 71–79.

[23] J. C. Willems and C. I. Byrnes, "Global adaptive stabilization in the absence of information on the sign of the high frequency gain," *Sixth Intl. Conf. on the Analysis and Opt. of Systems*, Nice, 1984.

[24] B. Mårtensson, "The order of any stabilizing compensator is sufficient a priori information for adaptive stabilization," *Systems and Control Letters*, **6**(1985), 87–92.

[25] C. I. Byrnes and J. C. Willems, "Intelligent set-point control," in preparation.

[26] W. H. Bennett and K. DeJong, "Adaptive search techniques and the design of decentralized control systems," *Naval Research Lab Report*, in publication.

[27] S. Arimoto, S. Kawamura and F. Miyazaki, "Bettering operation of robots by learning," *J. of Robotic Systems*, **1–2**(1984), 123–140.

[28] M. Takegaki and S. Arimoto, "A new feedback method for dynamic control of manipulators," *J. Dyn. Syst.*, **102**(1981), 119–125.

[29] B. Mårtensson, Ph.D. Thesis, Lund Institute of Technology, April 1986.

[30] D. E. Koditschek, "Adaptive strategies for the control of natural motion," *Proc. of 24th CDC*, Ft. Lauderdale, 1985, 1405–1409.

[31] K.-K. D. Young, P. V. Kokotovic and V. I. Utkin, "A singular perturbation analysis of high-gain feedback systems," *IEEE Trans. Aut. Control*, **AC-22**(1977), 931–938.

[32] R. Marino, "Nonlinear compensation by high gain feedback," to appear.

[33] P. V. Kokotovic, "Applications of singular perturbation techniques to control problems," *SIAM Review* (November 1984).

[34] C. I. Byrnes, U. Helmke and A. S. Morse, "Necessary conditions in adaptive control," *Modelling, Identification and Robust Control* (C. I. Byrnes and A. Lindquist, eds.), North Holland, 1986.

Christopher J. Byrnes received his Ph.D. at the University of Massachusetts and has held positions at the University of Utah and Harvard University. He is currently Research Professor at Arizona State University. His mathematical interests include the design of intelligent control systems and the application of modern mathematical techniques to other problems arising in control theory.

GLOBAL OBSERVABILITY OF ERGODIC
TRANSLATIONS ON COMPACT GROUPS

LANCE DRAGER   CLYDE MARTIN

Department of Mathematics
Texas Tech University
Lubbock, Texas 79409

ABSTRACT

Harmonic analysis is used to show that an ergodic translation on a compact abelian group is observed by almost all continuous scalar functions.

Key Words: Observability, ergodic, harmonic analysis.

1. Introduction.

We consider in this paper the problem of determining the behavior of a dynamical system when we are only allowed to observe some scalar function of the system. Let M be a manifold and let f be a vector field defined on M and we assume that we have for every x in M an integral curve of f initiating from x parametrized by time t. The theory of dynamical systems, of course, studies such systems in great detail. However in system theory there is an added complication—the manifold M is only observed through a real-valued function h. The problem which we consider in this paper is to determine what properties must be imposed on the vector field f, the manifold M and the function h in order that orbits of f are distinguished by h. So using local notation we consider the following definition.

Definition. *Let M, f, and h be as above and let x_0 and y_0 be any two points on M. Let $x(t)$ and $y(t)$ be the solutions of the differential equation $(d/dt)w(t) = f\big(w(t)\big)$ with initial data x_0 and y_0 respectively. The system determined by f and h is said to be observable iff $h\big(x(t)\big) = h\big(y(t)\big)$ implies that $x_0 = y_0$.*

Similar definitions exist in discrete time and the problem of observability has been studied in a variety of conditions. For example the condition is relevant when

37

M is a vector space over a finite field and f is a polynomial mapping from M to M and h is a mapping from the vector space into the finite field. In this paper we will find necessary conditions for a system to be observable when M is a compact group. The problem arose in the context of trying to determine when the winding lines on the n-torus are distinguished by real-valued functions. This problem has been studied by the authors in [4].

The results of this paper were presented by the second author at this conference. We dedicate this paper to Gail Young who gave the encouragement to do applied mathematics with the tools of modern analysis.

2. Local versus Global Problems.

The original observability problem of system theory was the problem of observing a linear system. Let A be an $n \times n$ matrix and let c be a $1 \times n$ matrix. Our system consists of the differential equation $(d/dt)x(t) = Ax(t)$ and the scalar function $y = cx$. Each initial condition x_0 gives a curve $y(t) = c \exp(At)x_0$. Linearity reduces the question of observability to the problem of distinguishing nonzero initial data from zero data. Thus in the linear case the system is observable iff the vanishing of Taylor coefficients of y implies that $x_0 = 0$. An easy and standard calculation shows that the system is observable iff the vectors $c, cA, cA^2, \ldots$ form a basis for R^n. In the nonlinear case this argument is doomed to failure for a variety of reasons. The system need not be analytic and even if it is the Taylor series may only converge in some small neighborhood of 0. This argument has been used to produce local conditions for observability. That is, if the system is sufficiently smooth then there exists a neighborhood of x_0 such that distinct Taylor coefficients imply observability.

For most classical problems in control theory the local theory sufficed, for almost all analysis was only concerned with the problem of controlling and observing systems around some operating point. However there is now a large set of problems in which it is necessary to understand the problem of observability from a global point of view. Global problems arose when the problems of system theory became more complex. For example in the design of control systems that will control an airplane through complicated maneuvers the relevant dynamical system evolves on a manifold that is the product of Euclidean space, tori and orthogonal groups.

In order to understand these problems the authors began a study in [3] and [4] to analyze the behavior of dynamical systems on the torus and to develop conditions under which such systems are observable. The spirit under which this is developing is that if the dynamics are sufficiently complicated almost any function will distinguish orbits. At this point we have found that ergodicity is a convenient tool but realize that other conditions could suffice equally well.

Global problems of control and observability have been studied in the control theory literature. Among the more notable is the work of Levitt and Sussman [9],

Aeyels [1], and Sontag [11]. In two recent papers by the authors, [3] and [4], global results have been obtained by using various complexity results. In [3] various properties of piecewise linear maps of the interval were used to prove global observability, and in [4] the approximation theorems of Kronecker were used to prove that the irrational translations on the n-torus Tor^n are observed by a large class of continuous functions.

3. Compact Topological Groups.

In this note, we will generalize from Tor^n to an arbitrary compact, metrizable, abelian group and show that an ergodic translation on such a group is observed by almost all continuous scalar-valued functions. We will use harmonic analysis, rather than the geometric approach of [4], so the present results give another approach to the observability of the ergodic translation flows on Tor^n.

To set up our machinery, let G be a compact, metrizable topological group, for the moment not necessarily abelian. Let μ denote Haar measure on G, normalized by $\mu(G) = 1$.

Let $a \in G$ be a fixed element and let $T : G \to G$ be translation by a, $T(g) = ag$. We will study the discrete time dynamical system generated by T.

Let $h : G \to \mathbb{C}$ be a complex-valued observation function. If we take an initial condition $g \in G$ and follow it forward under the dynamical system we get a sequence of observations $\{y_n(g)\}_{n=0}^{\infty}$ by $y_n(g) = h(T^n(g))$. h is said to observe T (in forward time) if distinct initial conditions give distinct sequences of observations. Equivalently, h observes T if $h(T^n(g_1)) = h(T^n(g_2))$ for all $n \geq 0$ implies that $g_1 = g_2$.

We will assume that the translation T is ergodic. This implies that G is necessarily abelian, [2, Chap. 4].

We will use some elementary machinery from harmonic analysis on locally compact abelian groups. Some standard references for this material are [10], [6], [8]. Recall that a character of G is a continuous group homomorphism from G to Tor^1, the group of complex numbers of modulus one. The set of characters forms a countable (discrete) group $\hat{G}$ under pointwise multiplication and the characters form an orthonormal basis for $L^2(G)$. The trivial character $\hat{e} = 1$ is the identity of $\hat{G}$. Thus, if $f \in L^2(G)$, we can form its Fourier series

$$f = \sum_{\chi \in \hat{G}} C_\chi(F)\chi,$$

$$C_\chi(f) = \langle \chi, f \rangle = \int \bar{\chi}(g) f(g) \, d\mu(g).$$

The series converges to f in L^2 norm. We will use the following standard fact about Fourier series. Let $b \in G$, and suppose that $f \in L^2(G)$ has the Fourier series $\sum a_\chi \chi$. If we define the translate of f by $(\tau_b f)(g) = f(bg)$, then the Fourier series of $\tau_b f$ is

$\sum a_\chi \chi(b)\chi$, i.e., $C_\chi(\tau_b f) = \chi(b)C_\chi(f)$. This is easily verified by changing variables in the integral for $\langle \chi, \tau_b f \rangle$.

If $g \in G$, let $O(g) = \{T^n(g) \mid n \in \mathbf{Z}\}$ be the orbit of g and let $O^+(g) = \{T^n(g) \mid n \geq 0\}$ be the forward orbit. The identity of G will be denoted by e.

4. Ergodic Flows and Observability.

The following proposition from [2, Chap. 4, Sect. 1] gives alternative characterizations of the ergodicity of T. The conditions (2) and (3) of this proposition were first studied by Halmos and Samelson in the classical paper [7].

Proposition 4.1. *The following conditions are equivalent.*

(1) The translation T is ergodic.

(2) $O(e) = \{a^n \mid n \in \mathbf{Z}\}$ is dense in G.

(3) $\chi(A) \neq 1$ for every non-trivial $\chi \in \hat{G}$.

Note that $g'O(g) = O(g'g)$, so if one orbit is dense, all orbits are dense.

To get observability in forward time, we will need the following lemma.

Lemma 4.2. $O(e) = \{a^n \mid n \in \mathbf{Z}\}$ *is dense if and only if* $O^+(e) = \{a^n \mid n \geq 0\}$ *is dense.*

To prove the non-trivial part of the lemma, assume that $O(e)$ is dense. We make the following claim.

Claim. *If U is a neighborhood of e and $N \geq 0$, there is a $k > N$ so that $a^k \in U$.*

If a is a torsion element (i.e., $a^n = e$ for some n), this is trivial, so assume that a is not a torsion element. In this case, the powers of a are distinct. We can find an open symmetric neighborhood V of e (i.e., $V^{-1} = V$) contained in the neighborhood $U \setminus \{a^n \mid 0 < |n| \leq N\}$. Since $V \setminus \{e\}$ is open, we can find some $a^\ell \in V \setminus \{e\}$. Since $a^\ell \notin \{a^n \mid 0 \leq |n| \leq N\}$, we have $|\ell| > N$. If $\ell > 0$, we are done and if $\ell < 0$, $a^{-\ell} \in V \subset U$, since V is symmetric. This proves the claim. To prove the lemma, let $g \in G$ and let U be a neighborhood of g. We must find a positive power of a in U. By continuity of multiplication, there are neighborhoods W and V, $g \in W$, and $e \in V$, so that $W \cdot V \subset U$. Since $O(e)$ is dense we can find some $a^k \in W$ and by the claim we can find $a^n \in V$ with $n > |k|$. Then $a^{k+n} \in U$ and $k + n > 0$. This completes the proof of the lemma.

We are now ready to state our main result. Let $C(G)$ be the space of continuous complex-valued functions on G. $C(G)$ is a Banach space in the supremum norm and, since G is compact, $C(G) \subset L^2(G)$. Define

$$R = \{f \in C(G) \mid C_\chi(f) \neq 0 \text{ for all } X \in \hat{G}\}.$$

Our main result is

Theorem 4.3.

(1) If T is an ergodic translation on G, T is observed (in forward time) by every function in R.

(2) Almost all continuous functions are in R, i.e., R is a residual subset of $C(G)$.

To prove (1), suppose that T is ergodic, so $O^+(e)$ is dense in G. Let $h \in R$ and suppose that for $g_1, g_2 \in G$ we have

(i) $h\big(T^n(g_1)\big) = h\big(T^n(g_2)\big)$ for $n \geq 0$.

We must show that $g_1 = g_2$. We may rewrite (i) as

(ii) $h(a^n g_1) = h(a^n g_2)$ for $n \geq 0$.

Write the Fourier series of h as $h = \sum a_\chi \chi$, with $a_\chi \neq 0$ for all $\chi \in \hat{G}$. Define $f : G \to \mathbb{C}$ by

$$f(g) = h(gg_1) - h(gg_2).$$

From the remarks above we see that the Fourier series of f is

$$f = \sum [a_\chi \chi(g_1) - a_\chi \chi(g_2)] \chi.$$

Now, (ii) says that $f = 0$ on $O^+(e)$. Since f is continuous and $O^+(e)$ is dense, we conclude $f \equiv 0$. But, if $f \equiv 0$, all the Fourier coefficients of f are zero, so

$$a_\chi \chi(g_1) = a_\chi \chi(g_2)$$

for all $\chi \in \hat{G}$. Since $a_\chi \neq 0$, we conclude

$$\chi(g_1) = \chi(g_2)$$

for all $\chi \in \hat{G}$. Since the character group $\hat{G}$ separates points, $g_1 = g_2$ and the proof of (1) is complete.

Note that we can actually conclude a bit more. If $S \subset \hat{G}$ is a set of characters that separates points and $R_S = \{f \in G(C) \mid C_\chi(f) \neq 0, \ \chi \in S\}$, then an ergodic translation is observed by all functions in R_S.

The proof of (2), is of course straightforward. Since G is compact, $C(G) \subset L^2(G)$ and the inclusion is continuous. Thus $f \mapsto C_\chi(f) = \langle \chi, f \rangle$ is a non-trivial continuous linear functional on $C(G)$ (non-trivial since $\chi \in C(G)$ and $C_\chi(\chi) = 1$). Thus the kernel K_χ of this linear functional is a closed subspace of codimension one and the complement K_χ^c of K_χ is a dense open set in $C(G)$. Then $R = \bigcap \{K_\chi^c \mid \chi \in \hat{G}\}$ is a countable intersection of dense open sets and so is by definition residual. By the Baire category theorem, R is dense in $C(G)$.

It is easy to see that both conclusions of the theorem hold if we restrict attention to continuous real-valued functions, since a real-valued function f has a Fourier series, subject to $C_{\bar{\chi}}(f) = C_\chi(f)^-$.

In closing, we remark that the arguments above go through equally well in the continuous time case. To see what is involved let $t \mapsto g_t$ be a continuous one parameter subgroup of G and consider the continuous time dynamical system $\phi_t(g) = g_t g$. The first observation to make here is that if $\chi \in \hat{G}$, the function $\Lambda_\chi(t) = \chi(g_t)$ is a continuous character of the additive group of the real numbers so $\chi(g_t) = \Lambda_\chi(t) = \exp(2\pi i \lambda_\chi t)$ for some $\lambda_\chi \in \mathbb{R}$. From [2], we have the following characterizations of the ergodicity of the flow $\{\phi_t\}$.

Proposition 4.4. *The following conditions are equivalent.*

(1) The flow $\{\phi_t\}$ is ergodic.

(2) $O(e) = \{g_t \mid t \in \mathbb{R}\}$ is dense.

(3) $\lambda_\chi = \dfrac{d}{dt}\chi(g_t)\Big|_{t=0} \neq 0$ for every non-trivial $\chi \in \hat{G}$.

All of the arguments above go through, with only trivial changes, to show that an ergodic translation flow $\{\phi_t\}$ is observed (in forward time) by any function in R.

In fact, we can apply the discrete time theorem directly to the continuous time case. Let $\{\phi_t\}$ be an ergodic translation flow, as above. By the proposition above, we have $\Lambda_\chi \not\equiv 1$ if $\chi \neq \hat{e}$, so $\operatorname{Ker} \Lambda_\chi$ is a countable discrete subset of $\mathbb{R}$. Since $\hat{G}$ is countable, the complement of $\bigcup\{\ker \Lambda_\chi \mid \chi \in \hat{G}, \ \chi \neq \hat{e}\}$ is dense in $\mathbf{R}$, so we can find $t_0 > 0$ so that $\chi(g_{t_0}) \neq 1$ for all non-trivial $\chi \in \hat{G}$. This means that the discrete time system T generated by g_{t_0} is ergodic, and so is observed in forward time by every function in R. But a function that observes this discrete time system certainly observes the continuous time system.

The authors would like to thank Professor Samelson for pointing out the reference to [7].

Acknowledgment. This research was supported in part by Air Force Contract No. F29601-85-C-0040, by NSA Grant No. MDA904-85-H-0009 and by NASA Grant No. NAG2-203.

References.

[1] D. Aeyels, "Global observability of Morse-Smale vector fields," *J. Diff. Eq.,* **45**(1982), 1–15.

[2] I. Cornfeld, S. Fomin and Y. Sianai, *Ergodic Theory,* Springer Verlag, Berlin, 1982.

[3] L. Drager and C. Martin, "Global observability of a class of nonlinear discrete time systems," *System and Control Letters,* **6**(1985), 65–68.

[4] L. Drager and C. Martin, "Global observability of flows on the torus; an application of number theory," to appear.

[5] R. Hermann and A. J. Krener, "Nonlinear controllability and observability," *IEEE Trans. Auto. Control,* **AC-22**(1977), 728–740.

[6] E. Hewitt and K. Ross, *Abstract Harmonic Analysis,* 2 vols., Springer Verlag, Berlin, 1963.

[7] P. R. Halmos and H. Samelson, "On monothetic groups," *Proc. Nat. Acad. Sci. U.S.A.*, **28**(1942), 254–258.

[8] Y. Katznelson, *An Introduction to Harmonic Analysis*, John Wiley and Sons, New York, 1968.

[9] N. Levitt and H. Sussman, "On controllability by means of two vector fields," *SIAM J. Cont. and Opt.*, **13**(1975), 1271–1281.

[10] W. Rudin, *Fourier Analysis on Groups*, Interscience, New York, 1962.

[11] E. Sontag, *Polynomial Response Maps*, Lecture Notes in Control and Information Sciences, Springer Verlag, 1979.

Lance Drager did his undergraduate work at the University of Minnesota, received his Ph.D. from Brandeis University, was a visiting member at the Courant Institute, and has taught at St. Francis College and the Georgia Institute of Technology. He is currently at Texas Tech University. His mathematical interests are in the applications of geometric techniques to differential equations and various applied problems.

Clyde Martin graduated from the University of Wyoming in 1971. Since then he has been at several universities and with NASA. He is currently Professor of Mathematics at Texas Tech University. His current interests are in the applications of mathematics to problems arising in agriculture, medical imaging, control theory and, most recently, in the management of migrating water fowl.

MATHEMATICAL MODELING AND LARGE-SCALE COMPUTING IN ENERGY AND ENVIRONMENTAL RESEARCH

RICHARD E. EWING

Departments of Mathematics and Petroleum Engineering
University of Wyoming
Laramie, Wyoming 82071

The author would like to dedicate this paper
to his friend, colleague, and mentor,
Gail Young

ABSTRACT

The advent of new super computer architectures is helping to revolutionize the modeling process for large-scale physical problems. The advance in computational capabilities has allowed the incorporation of more physics in the model. This has greatly increased the complexity of the mathematical models necessitating the application of a wider range of mathematical research. Through this development, the distinction between much of pure and applied mathematics is rapidly disappearing. The breadth of the modeling process, utilizing physical, mathematical, numerical, and computational concepts, will be emphasized. Specific applications involving energy-related and environmental problems will be presented and used to illustrate the need for strong interrelationships between the aspects of the steps in modeling.

1. Introduction.

The need for the study and use of mathematics is growing and expanding extremely rapidly in response to the enormous recent development of computing capabilities. The use of complex models which incorporate more detailed physics has necessitated more sophisticated mathematics in the modeling process. In this way a broader range of mathematics is needed for applications and the divisions between pure and applied

45

mathematics are disappearing, to the benefit of all of mathematics. In this paper, we shall discuss some aspects of the expanding scope of mathematical modeling.

The mathematical techniques which are used to model energy-related and environmental problems are representative of those needed for many other applications and will be used to illustrate the role of mathematics in modeling. The need to understand the various phenomena governing the energy-related applications and to optimize the production of energy resources while monitoring and controlling associated environmental problems is becoming increasingly important. The advent of orders-of-magnitude better computing capabilities has allowed the modeling of more complicated physical phenomena. We will indicate how this growth is changing the entire modeling process. Modeling of large-scale physical processes involves four major interrelated stages. First, a physical model of the physical processes must be developed incorporating as much physics as is deemed necessary to describe the essential phenomena. A careful list of the assumptions made in establishing this physical model should be compiled. Second, a mathematical formulation of the physical model should be obtained, usually involving coupled systems of nonlinear partial differential equations. The properties of this mathematical model, such as existence, uniqueness, and regularity of the solution are then obtained and related to the physical process to check the model. Third, a discretized numerical model of the mathematical equations is produced. This numerical model must have the properties of accuracy and stability and produce solutions which represent the basic physical features as well as possible without introducing spurious phenomena associated with the specific numerical schemes. Fourth, a computer program capable of efficiently performing the necessary computations for the numerical model is sought. Although the total modeling process encompasses aspects of each of these four intermediate stages, the process is not complete with one pass through the steps. Usually many iterations of this modeling loop are necessary to obtain reasonable models for the highly complex physical phenomena involved in energy and environmental topics.

The aims of this paper are to introduce certain complex physical phenomena which need to be better understood, to illustrate aspects of the modeling process used to describe these processes, and to discuss some of the newer mathematical tools that are being utilized in the various models. The complexity of the models requires sophisticated mathematical analysis. For example, the increasing use of large, coupled systems of nonlinear partial differential equations to describe the movement of multiphase and multicomponent fluid systems through porous media is identifying very difficult problems in the theoretical aspects of the partial differential equations, the numerical analysis of various discretization schemes, the development of new, accurate numerical models, and the computational efficiency of discrete systems resulting from the discretizations. The interplay between the engineering and physics of the applications, the mathematical properties of the models and discretizations, and the role of the computer in the algorithm development is critical and will be stressed in this presentation.

The modeling of many fluid flow problems involves very similar mathematical equations. Examples of mathematical and related physical properties of these models which must be addressed include: (a) the resolution of sharp moving fronts in convection dominated convection-diffusion problems, (b) the stability and accuracy of discretization of highly non-self-adjoint differential operators, (c) the need to have very accurate fluid velocities which dominate the flow, (d) the need to model dynamic local phenomena which govern the physics, and (e) the emphasis on development of efficient numerical procedures for the enormous problems encountered.

A model problem which illustrates many major numerical difficulties arising in reservoir simulation is presented in Section 2. The numerical stability problems associated with this transport dominated system and the corresponding pure transport problem are discussed. A modified method of characteristics based on combining the transport and accumulation terms in the equation into a directional derivative along characteristic-like curves is then briefly described. The modified method of characteristics is heavily dependent upon having very accurate fluid velocities. Section 3 is then devoted to the description of a mixed finite element procedure which is designed to give approximations of the fluid velocities which are just as accurate as the pressure approximations, even in the context of rapidly changing reservoir properties. The need for adaptive local grid refinement methods to resolve certain dynamic, highly localized physical phenomena is described in Section 4. Important considerations such as a choice of versatile and efficient data structures and adaptivity techniques are discussed.

2. Description of Model Problem and Modified Method of Characteristics.

A model system of equations describing the miscible displacement of one incompressible fluid by another in a thin horizontal porous medium is given by [20,31,39]

$$\nabla \cdot \mathbf{u} = -\nabla \cdot \frac{k}{\mu(c)} \nabla p = q, \qquad \mathbf{x} \in \Omega, \quad t \in J \qquad (1)$$

$$\phi \frac{\partial c}{\partial t} + \nabla \cdot [\mathbf{u} c - D(\mathbf{u}) \nabla c] = \bar{c} q, \qquad \mathbf{x} \in \Omega, \quad t \in J \qquad (2)$$

$$\mathbf{u} \cdot \mathbf{n} = [\mathbf{u} c - D(\mathbf{u}) \nabla c] \cdot \mathbf{n} = 0, \qquad \mathbf{x} \in \partial\Omega, \, t \in J \qquad (3)$$

$$c(\mathbf{x}, 0) = c_0(x), \qquad \mathbf{x} \in \Omega \qquad (4)$$

for $\Omega \in \mathbb{R}^2$ with boundary $\partial\Omega$ and $J = [0, T]$, where p and $\mathbf{u}$ are the pressure and velocity of the single phase fluid mixture, c is the concentration of the injected fluid, and q is the total volumetric flow rate, modeled by Dirac delta point sources and sinks describing the injection and production wells, ϕ, k, μ, and $D(\mathbf{u})$ are assumed to be known rock and fluid properties. D is a diffusion-dispersion tensor given by [20,28,39]

$$\left(D_{ij}(\mathbf{x}, \mathbf{u})\right) = \phi d_m I + \frac{d_l}{|\mathbf{u}|} \begin{pmatrix} u_1^2 & u_1 u_2 \\ u_1 u_2 & u_2^2 \end{pmatrix} + \frac{d_t}{|\mathbf{u}|} \begin{pmatrix} u_2^2 & -u_1 u_2 \\ -u_1 u_2 & u_1^2 \end{pmatrix} \tag{5}$$

where $\mathbf{u} = (u_1, u_2)$, $|\mathbf{u}|$ is the Euclidean norm of $\mathbf{u}$, d_m is the molecular diffusion coefficient, and d_l and d_t are the magnitudes of longitudinal and transverse dispersion. Equation (2) is an example of a transport dominated convection-diffusion equation. Since diffusion is small, the solution c exhibits very sharp fronts or concentration gradients which move in time across the reservoir and finger into production wells. The frontal width is very narrow in general, but must be resolved accurately via the numerical method since it describes the physics of the mixing zone and governs the speed of the frontal movement, and thus the production history of the hydrocarbons. Similar dispersive mixing zones are critical in the modeling of contaminant transport processes.

If the dispersion tensor in Equation (2) is ignored as in most simulators in use today, Equation (2) becomes a first order hyperbolic problem instead of a transport dominated convection-diffusion equation. Standard highly accurate finite difference schemes for hyperbolic partial differential equations are known to be unstable and various upstream weighting or "artificial diffusion" techniques have been utilized to stabilize the variant of Equation (2). The upstream weighting techniques (described in [20]) used in the petroleum industry introduce artificial diffusion in the direction of the grid axes and of a size proportional to the grid spacings. Thus, although this stabilizing effect would be small if very fine grid block spacings were used, the enormous size of petroleum simulation problems necessitates the use of large grid blocks and hence, large, directionally-dependent artificially induced numerical diffusion which has nothing to do with the physics of the flow. Two major problems in numerical reservoir simulation today are due essentially to the use of standard upstream weighting techniques. First, the upstream methods, by introducing a large artificial numerical diffusion or dispersion, smear sharp fluid interfaces producing erroneous predictions of the degree of mixing and incorrect frontal velocities. Second, the numerical diffusion is generated along grid lines and produces results which are radically different if the orientation of the grid is rotated forty-five degrees. This "grid orientation problem" and several attempts to alleviate it are described in more detail in [20].

The use of physical intuition in determining a more accurate numerical scheme can be illustrated in this case. The physical diffusion-dispersion term displayed in Equation (5) is a rotationally-invariant tensor. Therefore, one way to stabilize the first order hyperbolic problem without introducing artificial directional effects is to use an "artificial diffusion" term of the form in Equation (5). The size of this term must then be closely considered in order not to diffuse fronts too badly. A consequence of this type of stabilization with finite difference discretization means a nine-point difference star would be necessary to approximate the cross-derivatives accurately instead of the standard five-point star used in two space dimensions. In three space dimensions a twenty-seven point star would be necessary to replace a seven point star. If iterative solution techniques are being utilized this greatly increases the solution times. This is a good example of how decisions made in one part of the modeling process can greatly influence other parts of the problem.

For more complex physical processes, the system of Equations (1)–(4) must be expanded to include mass balances from different components or different phases. The governing equations for in-situ combustion processes, for example, could involve coupled systems of up to twenty nonlinear partial differential equations of the form of Equation (2). The interaction between these coupled nonlinear equations can greatly affect the properties of the equations. Much work must be done to understand the mathematical properties of existence, uniqueness, and continuous dependence of solutions upon data for coupled systems of this form. Therefore, the improved computing capabilities which allow the numerical approximation of large, coupled systems of nonlinear partial differential equations, are necessitating the theoretical study of properties of systems of these equations. The "applied" mathematician involved in the simulation must understand and be able to work with these "purer" areas if the modeling process is to be effective.

The numerical analysis involved in rigorously obtaining asymptotic error estimates for even the model problem presented in Equations (1)–(5) requires various aspects of functional analysis and approximation theory. The order of the approximations depends upon the use of fractional order Sobolev spaces and interpolation spaces. Asymptotic error estimates for the miscible displacement problem appear in [28,30,31,32].

Russell described a technique based on a method of characteristics approach for treating the first order hyperbolic part of Equation (2). This technique [19,37], based on a form of Equation (2) which is analogous to a convection-diffusion equation, was implemented by Russell [37,38] and now forms the basis for our time-stepping scheme.

In order to introduce a nondivergence form of Equation (2) that is used in our numerical schemes, we first expand the convection $(\nabla \cdot \mathbf{u}c)$ term with the product rule and use Equation (1) to obtain

$$\phi \frac{\partial c}{\partial t} + \mathbf{u} \cdot \nabla c - \nabla \cdot [D(\mathbf{u})\nabla c] = (\bar{c} - c)\bar{q}, \qquad \mathbf{x} \in \Omega,\ t \in J \qquad (6)$$

where $\bar{q} = \max\{q,0\}$ is nonzero at injection wells only. To avoid technical boundary difficulties associated with our modified method of characteristics for Equation (6), we assume that Ω is a rectangle and that the problem given by Equations (1), (6), (3), and (4) is Ω-periodic.

The basic idea is to consider the hyperbolic part of Equation (6), namely, $\phi \partial c / \partial t + \mathbf{u} \cdot \nabla c$, as a directional derivative. Accordingly, let $\mathbf{s}$ denote the unit vector in the direction of (u_1, u_2, ϕ) in $\Omega \times J$, and set

$$\psi(\mathbf{x}) = \left(u_1(\mathbf{x})^2 + u_2(\mathbf{x})^2 + \phi^2 \right)^{1/2}. \tag{7}$$

Then Equation (6) can be rewritten in the form

$$\psi \frac{\partial c}{\partial \mathbf{s}} - \nabla \cdot (D \nabla c) + \bar{q} c = \bar{q} \, \bar{c}. \tag{8}$$

Note that the spatial operator in Equation (8) is now self-adjoint, symmetric matrices will result from spatial discretization, and the associated numerical methods will be better behaved. Since iterative solution techniques are used to solve the nonlinear equations resulting from finite element discretization of Equation (8), and since symmetry is very important in any of the useful conjugate gradient iterative solvers, this change to symmetric matrices is very important.

One critical aspect of the modified method of characteristics is the need for accurate approximation of the directional derivative $\partial c / \partial \mathbf{s}$. Many methods based upon characteristics fix a grid at time t^{n-1} and try to determine where these points would move under the action of the characteristics. These "moving point" or "front tracking" methods must then discretize Equation (6) and solve for the unknowns c^n on a mesh of irregular or unpredictable nature. If too large a time-step is chosen, serious difficulties can arise from the spatial and temporal behavior of the characteristics. Front-tracking in two space dimensions is difficult while in three dimensions, it is considerably more difficult. For details of the discretization of $\partial c / \partial \mathbf{s}$ and the ideas for extending this method to higher space dimensions, see [29,30].

3. Mixed Finite Elements for Pressure and Velocity.

Since both the modified method of characteristics and the diffusion-dispersion term in Equation (6) are governed by the fluid velocity, accurate simulation requires an accurate approximation of the velocity $\mathbf{u}$. In order to use large grid blocks for simulation and still capture the effects of a very small well bore (relative to the grid block size), the near-singular behavior of the pressure and fluid velocities "near" the wells are often modeled via singular functions "at" the wells. Using the Dirac delta functions for well models, we see that $\mathbf{u}$ is not even square integrable at the wells and thus standard approximations for $\mathbf{u}$ do not converge at the wells. A technique for removing the leading singularity in $\mathbf{u}$ and accurately approximating the result is presented. Also, since the lithology in the reservoir can change abruptly, causing rapid changes in the flow capabilities of the rock, the coefficient k in Equation (1) can be discontinuous. In this case, in order for the flow to remain relatively smooth, the pressure changes extremely rapidly. Thus standard procedures of solving Equation (1) as an elliptic partial differential equation for pressure, differentiating or differencing the result to approximate the pressure gradient, and then multiplying by the discontinuous k/μ can produce very poor approximations to the velocity $\mathbf{u}$. In this section a mixed finite element method for approximating $\mathbf{u}$ and p simultaneously, via a coupled system of first order partial differential equations, will be discussed. This formulation allows the removal of singular terms in the equations and accurately treats the problem of rapidly changing flow properties in the reservoir.

The coupled system of first order equations used to define our methods arise from Darcy's Law and conservation of mass

$$\mathbf{u} = -\frac{k}{\mu}\nabla p, \qquad\qquad \mathbf{x} \in \Omega, \qquad (9)$$

$$\nabla \cdot \mathbf{u} = q, \qquad\qquad \mathbf{x} \in \Omega, \qquad (10)$$

subject to the boundary condition

$$\mathbf{u} \cdot \mathbf{n} = 0, \qquad\qquad \mathbf{x} \in \partial\Omega. \qquad (11)$$

Clearly Equations (9)–(11) will determine p only to within an additive constant. Thus a normalizing constraint such as $\int_\Omega p(\mathbf{x})\,d\mathbf{x} = 0$ or $p(\mathbf{x}_s) = 0$ for some $\mathbf{x}_s \in \Omega$ is required in the computation to prevent a singular system.

We next define certain function spaces and notation. Let $W = L^2(\Omega)$ be the set of all functions on Ω whose square is finite integrable. Let $H(\mathbf{div};\Omega)$ be the set of vector functions $\mathbf{v} \in [L^2(\Omega)]^2$ such that such $\nabla \cdot \mathbf{v} \in L^2(\Omega)$ and let

$$V = H(\mathbf{div};\Omega) \cap \{\mathbf{v} \cdot \mathbf{n} = 0 \quad \text{on } \partial\Omega\}. \qquad (12)$$

Let $(v,w) = \int_\Omega vw\,dx$, $\langle v,w \rangle = \int_{\partial\Omega} wv\,ds$, and $\|v\|^2 = (v,v)$ be the standard L^2 inner products and norm on Ω and $\partial\Omega$. We obtain the weak solution form of Equations (9)–(11) by dividing each side of Equation (9) by k/μ, multiplying by a test function $\mathbf{v} \in V$, and integrating the result to obtain

$$\left(\frac{\mu}{k}\mathbf{u}, \mathbf{v}\right) = (p, \nabla \mathbf{v}), \qquad\qquad \mathbf{v} \in V. \qquad (13)$$

The right-hand side of Equation (13) was obtained by further integration by parts and use of Equation (12). Next, multiplying Equation (10) by $w \in W$ and integrating the result, we complete our weak formulation, obtaining

$$(\nabla \cdot \mathbf{u}, w) = (q, w) \qquad\qquad w \in W. \qquad (14)$$

For a sequence of mesh parameters $h > 0$, we choose finite dimensional subspaces V_h and W_h with $V_h \subset V$ and $W_h \subset W$ and seek a solution pair $(\mathbf{U}_h; P_h) \in V_h \times W_h$ satisfying

$$\left(\frac{\mu}{k}\mathbf{U}_h, \mathbf{v}_h\right) - (P_h, \mathbf{div}\,\mathbf{v}_h) = 0, \qquad\qquad \mathbf{v}_h \in V_h, \qquad (15)$$

$$(\text{div } \mathbf{U}_h, w_h) = (q, w_h), \qquad\qquad w_h \in W_h. \qquad (16)$$

We can now complete the description of our mixed finite element methods with a discussion of particular choices of V_h and W_h. Examples of these spaces are presented in [33].

For problems with smooth coefficients and smooth forcing functions, standard approximation theory results show that, by using higher order basis functions, correspondingly higher order convergence rates can be obtained. For the fluid flow in porous media applications, the source and sink terms q are not smoothly distributed, but are sums of Dirac delta functions. As shown in [32,33], the resulting smoothness of $\mathbf{u}$ is reduced; $\mathbf{u}$ is not contained in the space L^2 and thus using the methods described by Equations (15) and (16), the velocity approximations $\mathbf{U}_h$ *would not converge at the wells*. This result was obtained theoretically in [17,18,32] and computationally in [26]. By removing the leading term of the singularities (the logarithm terms), the remaining parts of the velocities are now in $H^{2-\epsilon}$ for any $\epsilon > 0$. Thus the approximations to these parts will now converge at the wells since we have regained sufficient regularity for convergence.

In order to treat the point sources and sinks which model wells in our codes, we remove the singularities at the wells for the velocities and then solve for the remaining portions via the mixed finite element techniques described above. We decompose $\mathbf{u}$ into its regular and singular parts ($\mathbf{u}_r$ and $\mathbf{u}_s$, respectively):

$$\mathbf{u} = \mathbf{u}_r + \mathbf{u}_s, \qquad\qquad (17)$$

$$\mathbf{u}_s = \sum_{j=1}^{N_w} Q_j(t) \nabla N_j, \qquad\qquad (18)$$

$$N_j = \frac{1}{2\pi} \log |\mathbf{x} - \mathbf{x}_j|, \qquad j = 1, 2, \ldots, N_w, \qquad (19)$$

where N_w is the number of wells, $Q_j(t)$ are the flow rates at the wells located at $\mathbf{x}_j$, and $\mathbf{u}_r$, the regular part of $\mathbf{u}$, satisfies the relations

$$\nabla \cdot \mathbf{u}_r = 0, \qquad\qquad \mathbf{x} \in \Omega, \qquad (20)$$

$$\mathbf{u}_r \cdot \mathbf{n} = -\mathbf{u}_s \cdot \mathbf{n}, \qquad\qquad \mathbf{x} \in \partial\Omega. \qquad (21)$$

Let $\mathbf{U}_r$ be the finite element approximation to $\mathbf{u}_r$ from V_h, let

$$\mathbf{U} = \mathbf{U}_r + \mathbf{u}_s \qquad\qquad (22)$$

be our numerical approximation of $\mathbf{u}$, and let $P_h \in W_h$ be our approximation to p. We then see that $\mathbf{U}_r \in V_h$ satisfies

$$\left(\frac{\mu}{k}\mathbf{U}_r, \mathbf{v}_h\right) - (P_h, \text{div } \mathbf{v}_h) = -\left(\frac{\mu}{k}\mathbf{u}_s, \mathbf{v}_h\right), \qquad\qquad \mathbf{v}_h \in V_h, \qquad (23)$$

$$(\mathbf{div}\,\mathbf{U}_r, w_h) = 0, \qquad\qquad w_h \in W_h, \qquad (24)$$

$$\langle(\mathbf{U}_r + \mathbf{u}_s) \cdot \mathbf{n}, \mathbf{v}_h \cdot \mathbf{n}\rangle = 0, \qquad\qquad \mathbf{v}_h \in V_h. \qquad (25)$$

We note that Equation (25) requires that the net flow across $\partial\Omega$ of each boundary element be zero.

Special choices of basis functions for the Raviart-Thomas spaces [35] based upon Gauss-point nodal functions and related quadrature rules have significantly aided in the computational efficiency of these methods. For detailed descriptions of these bases and computational results, see [14,26,32]. The observed convergence rates matched those predicted in [17,18]. Also superconvergence results were obtained at specific locations which can be utilized in quadrature and reduced quadrature considerations in the coupled systems described in Section 2.

Since the set of equations (9)–(11) will only determine the pressure to within an arbitrary constant, the algebraic system arising from our mixed method system (23)–(25) is not definite unless constants are modded out of the approximating space W_h for pressures. If the unknowns for the x and y components of the velocity are formally eliminated from the resulting system, one can obtain a set of equations for the pressure variable. The matrix arising in this problem is quite complex, but is comparable to a matrix generated by finite difference methods for the pressure [39]. Preconditioned conjugate gradient iterative procedures have been developed to efficiently solve this set of linear equations [26,41].

Techniques for coupling the mixed finite element procedures with a modified method of characteristics for the concentration in Equations (2)–(6) have appeared in the literature [23,29]. Asymptotic error estimates and convergence rates for this coupled procedure also appeared in [30]. We note that the incorporation of analytic techniques for singularity removal is easy to accomplish in a finite element setting where integrals can take advantage of the knowledge of the asymptotic behavior of the velocity near the wells for more accurate simulation.

For fluid flow around production wells in a procedure where the injected fluid is much less viscous than the resident fluid, a viscous fingering phenomenon dominates the flow, destroying radial flow characteristics. In this case local grid refinement around the production wells holds great promise. In the next section we will describe some recent advances in self-adaptive local grid refinement techniques for reservoir simulation applications.

4. Adaptive Local Grid Refinement.

Many of the chemical and physical phenomena which govern enhanced recovery processes or contaminant flow processes have extremely important local properties. Thus the models used in simulators for these problems must be capable of resolving these critical local features. Also, in order to be useful in large-scale dynamic simulators, these models must be self-adaptive and extremely efficient. The development of adaptive grid refinement techniques must take into account the rapid development of new, advanced computer architectures. The compatibility of adaptive mesh modification algorithms with the intended computer is a critical consideration in the algorithm development.

In many enhanced recovery processes, fluids are injected into some wells in a reservoir while the resident hydrocarbons are produced from other wells, designated production wells. As one fluid displaces the others the complex chemical and physical processes which occur along the moving interface between the fluids govern the effectiveness of the process. In order to model these important extremely localized phenomena, a grid spacing which is more nearly of the order of the process must be used. When the length scale of the physics is on the size of one or two feet or smaller, local grid refinement techniques are essential for large scale simulation. Also, since monitoring the solution processes and manually changing the grid when necessary would be impossible for large simulation problems, the grid refinement strategies must be self-adaptive.

In many enhanced recovery processes where the injected fluid is much less viscous than the resident fluids, a strong fingering phenomena is present near the production wells after the injected fluids have reached the production wells and the radial flow models discussed in Section 3 are not sufficiently accurate. In these cases local grid refinement around the wells is very useful and can greatly increase the accuracy of the simulation throughout the reservoir.

The flexibility to dynamically change the number of grid points and thus the number of unknowns can create difficulties in the linearization and linear solution algorithms. In particular, it is extremely difficult to vectorize codes with changing numbers of unknowns for efficient solution on vector machine architectures. The ability to have truly local refinement and derefinement capabilities necessitates the use of a fairly complex data structure. A data structure with these properties was developed at Mobil's Field Research Laboratory and has been described in the literature [15,21–24]. It is a multilinked list which utilizes various properties of the tree structures presented in [36] and [7,8]. The structure allows efficient linear solution algorithms via tree-traveling techniques. Although these algorithms are extremely difficult to vectorize effectively, the tree structure lends itself well to parallelism at many levels. Development of codes for efficient use of MIMD (Multiple Instruction Multiple Data Stream) architectures to parallelize the local grid refinement algorithms based upon a multiple linked-list data structure are under way [16]. Preliminary experience with the Denelcor HEP, an MIMD machine, in this context has been educational.

For truly general local refinement a complex data structure like those discussed above and associated complications to the code are necessary. If local refinement is only needed in a very few special points, a technique termed patch refinement may be an attractive alternative. These concepts do not require as complex a data structure but do involve ideas of passing information from one uniform grid to another. Berger and Oliger have been using patch refinement techniques for hyperbolic problems using finite difference discretizations for some time [9,10].

The idea of a local-patch refinement method is to pick a patch that includes most of the critical behavior around the singular point and do a much finer, uniform grid refinement within this patch. Given a uniform fine grid, very fast solvers can be applied locally in this region using boundary data from the coarse original grid. McCormick, Thomas and co-workers have used multigrid techniques to solve the fine-grid problem in a simple two-well pressure model [27]. They have addressed the communication problem with the coarse grid and have attained conservation of mass on their "composite grid." Extensions of their technique, termed FCOM (fast composite mesh method), to more difficult problems are planned.

Bramble, Pasciak and Schatz [10–12] have developed some efficient gridding and preconditioning techniques which can also be used in the local-patch refinement framework. Their methods have logically rectangular grids within the patch which can be solved very rapidly via FFT preconditioners. The important problem is the communication between grids. Recent work by Bramble, Pasciak, Schatz and Ewing [13] uses preconditioning for local grid refinement in a way to make implementation of the methods in existing large scale simulators an efficient process. These techniques, based upon finite element preconditioners, could help produce a major advancement in incorporating fixed local refinement methods in a wide variety of applications and existing codes. We also feel that these methods are sufficiently powerful to handle local time-stepping applications as well.

The adaptivity of the local refinement methods must be driven either by a type of "activity index," which relays rapid changes in solution properties, or by some estimate of the errors present in different spatial locations which need to be reduced. Recently, locally-computable a posteriori error estimators have been developed by Babuška and Rheinboldt [3–5], Bank [6], Weiser [40], and Oden [34]. Under suitable assumptions, these error estimators converge to the norm of the actual error as the mesh size tends to zero. These a posteriori error estimators are extremely important for problems involving elliptic partial differential equations in determining the reliability of estimates for a fixed grid and a fixed error tolerance in a given norm. The error estimators are used to successively refine locally until the errors in a specified norm are, in some sense, equilibrated. Although these methods are very effective for elliptic problems, they are not efficient for large time-dependent problems where an "optimal" mesh at each time step is not "optimal" for the entire time-dependent problem.

For hyperbolic or transport dominated parabolic partial differential equations, sharp fronts move along characteristic or near-characteristic directions. Therefore the computed velocity determines both the local speed and direction of the regions where local refinement will be needed at the next time steps. This information should be utilized to help move the local refinement with the front. Although patch refinement techniques based upon characteristic-direction adaptation strategies do not determine a "locally optimal" grid, the waste in using more grid than necessary is compensated for by the overall efficiency. Use of a larger refined area and grid movement only after several time-steps is the technique that we are developing since efficiency is crucial in large-scale reservoir simulation.

Variable coefficients in the partial differential equations significantly complicate local refinement techniques for finite difference methods. At present, techniques for weighting the finite difference stars based upon the varying coefficient values seem to be "ad hoc" and can often cause serious errors in the flow description. Local refinement techniques with finite element methods always yield a straightforward way to evaluate and weight the coefficients and are, in general, much easier to apply. Thus the versatility of variational techniques often more than compensates for the slight addition in computational complexity of finite element methods.

Acknowledgment. This research is supported in part by the U.S. Army Research Office Contract No. DAAG29-84-K-002, by the U.S. Air Force Office of Scientific Research Contract No. AFOSR-85-0117, and by the National Science Foundation Grant No. DMS-8504360.

References.
[1] M. B. Allen and R.E. Ewing, "Applied mathematics in groundwater hydrology and contaminant transport," *SIAM News,* **18**(1985), 3,14.
[2] M. B. Allen, R. E. Ewing and J. V. Koebbe, "Mixed finite-element methods for computing groundwater velocities," *Numerical Methods for Partial Differential Equations,* in press.
[3] I. Babuška and W. C. Rheinboldt, "A-posteriori error estimates for the finite element method," *Internat. J. Numer. Meths. Engrg.,* **12**(1978), 1597–1615.
[4] I. Babuška and W. C. Rheinboldt, "Error estimates for adaptive finite element computation," *SIAM J. Numer. Anal.,* **15**(1978), 736–754.
[5] I. Babuška and W. C. Rheinboldt, "Reliable error estimation and mesh adaption for the finite element method," in *Computational Methods in Nonlinear Mechanics* (J. T. Oden, ed.), North-Holland, New York, 1980.
[6] R. E. Bank, "A multi-level iterative method for nonlinear elliptic equations," in *Elliptic Problem Solvers* (M. Schultz, ed.), Academic Press, New York, 1981.
[7] R. E. Bank and A. H. Sherman, "PLTMG users' guide," *Technical Report No. 152,* University of Texas at Austin, Center for Num. Anal., 1979.
[8] R. E. Bank and A. H. Sherman, "A refinement algorithm and dynamic data structure for finite element meshes," *Technical Report No. 166,* University of Texas at Austin, Center for Num. Anal., 1980.
[9] M. J. Berger, "Data structures for adaptive mesh refinement," in *Adaptive Computational Methods for Partial Differential Equations* (I. Babuska, J. Chandra and J. E. Flaherty, eds.), SIAM, Philadelphia, 1983, 237–251.

[10] M. J. Berger and J. Oliger, "Adaptive mesh refinement for hyperbolic partial differential equations," *Man. NA-83-02*, Computer Science Dept., Stanford University, 1983.

[11] J. H. Bramble, J. E. Pasciak and A. H. Schatz, "An iterative method for elliptic problems on regions partitioned into substructures," *Math. Comput.*, in press.

[12] J. H. Bramble, J. E. Pasciak and A. H. Schatz, "The construction of preconditioners for elliptic problems by substructuring, I," *Math. Comput.*, in press.

[13] J. H. Bramble, J. E. Pasciak, A. H. Schatz and R. E. Ewing, "A preconditioning technique for the efficient solution of problems with local grid refinement," *Comp. Meth. Appl. Mech. Engrg.*, in press.

[14] B. L. Darlow, R. E. Ewing and M. F. Wheeler, "Mixed finite element methods for miscible displacement problems in porous media," SPE 10501, *Proc. Sixth SPE Symp. on Reservoir Simulation*, New Orleans, 1982, 137–145; *Soc. Pet. Engrs. J.*, 4(1984), 391–398.

[15] J. C. Diaz, R. E. Ewing, R. W. Jones, A. E. McDonald, I. M. Uhler and D. U. von Rosenberg, "Self-adaptive local grid refinement for time-dependent, two-dimensional simulation," in *Finite Elements in Fluids*, Vol. VI, Wiley, New York, 1984.

[16] J. C. Diaz and R. E. Ewing, "Potential of HEP-like MIMD architectures in self adaptive local grid refinement for accurate simulation of physical processes," *Proc. Workshop on Parallel Processing Using the HEP*, Norman, Oklahoma, 1985.

[17] J. Douglas Jr., R. E. Ewing and M. F. Wheeler, "The approximation of the pressure by a mixed method in the simulation of miscible displacement," *R.A.I.R.O. Analyse Numerique*, 17(1983), 17–34.

[18] J. Douglas Jr., R. E. Ewing and M. F. Wheeler, "Time-stepping procedures for simulation of miscible displacement using mixed methods for pressure approximation," *R.A.I.R.O Analyse Numerique*, 17(1983), 249–265.

[19] J. Douglas Jr. and T. F. Russell, "Numerical methods for convection dominated diffusion problems based on combining the method of characteristics with finite element or finite difference procedures," *SIAM J. Numer. Anal.*, 19(1982), 871–885.

[20] R. E. Ewing, "Problems arising in the modeling of processes for hydrocarbon recovery," in *Research Frontiers in Applied Mathematics*, Vol. 1 (R. E. Ewing, ed.), SIAM, Philadelphia, 1983.

[21] R. E. Ewing, ed., "Oil Reservoir Simulation," *Comput. Meths. Appl. Mech. Engrg.*, 47(1984) (special issue).

[22] R. E. Ewing, "Adaptive mesh refinement in petroleum reservoir simulation," in *Accuracy Estimates and Adaptivity for Finite Elements* (I. Babuska, O.C. Zienkiewicz and E. Arantes e Oliveira, eds.), Wiley, New York, 1985.

[23] R. E. Ewing, "Finite element methods for nonlinear flows in porous media," *Comput. Meths. Appl. Mech. Engrg.*, in press.

[24] R. E. Ewing, "Efficient adaptive procedures for fluid flow applications," *Comput. Meths. Appl. Mech. Engrg.*, in press.

[25] R. E. Ewing and J. V. Koebbe, "Mixed finite element methods for groundwater flow and contaminant transport," *Proc. Fifth IMACS Internat. Symp. on Partial Differential Equations*, Bethlehem, Pennsylvania, 1984.

[26] R. E. Ewing, J. V. Koebbe, R. Gonzalez and M. F. Wheeler, "Computing accurate velocities for fluid flow in porous media," *Proc. Fifth Internat. Symp. on Finite Elements and Flow Problems*, Austin, Texas, 1984, 233–249.

58

[27] R. E. Ewing, S. McCormick and J. Thomas, "The fast adaptive composite grid method for solving differential boundary-value problems," *Proc. Fifth ASCE Specialty Conf.*, Laramie, Wyoming, 1984, 1453–1456.

[28] R. E. Ewing, and T. F. Russell, "Efficient time-stepping methods for miscible displacement problems in porous media," *SIAM J. Numer. Anal.*, **19**(1982), 1–66.

[29] R. E. Ewing, T. F. Russell and M. F. Wheeler, "Simulation of miscible displacement using mixed methods and a modified method of characteristics," *Proc. Seventh SPE Symp. on Reservoir Simulation*, San Francisco, 1983.

[30] R. E. Ewing, T. F. Russell and M.F. Wheeler, "Convergence analysis of an approximation of miscible displacement in porous media by mixed finite elements and a modified method of characteristics," *Comput. Meth. Appl. Mech. Engrg.*, **47**(1984), 73–92.

[31] R. E. Ewing, and M.F. Wheeler, "Galerkin methods for miscible displacement problems in porous media," *SIAM J. Numer. Anal.*, **17**(1980), 351–365.

[32] R. E. Ewing and M. F. Wheeler, "Galerkin methods for miscible displacement problems with point sources and sinks — unit mobility ratio case," in *Lectures on the Numerical Solution of Partial Differential Equations*, University of Maryland, 1981, 151–174.

[33] R. E. Ewing and M. F. Wheeler, "Computational aspects of mixed finite element methods," in *Numerical Methods for Scientific Computing* (R.S. Stepleman, ed.), North-Holland Publishing Co., 1983, 163–172.

[34] J. T. Oden, "Adaptive methods for incompressible viscous flow with moving boundaries," in *Accuracy Estimates and Adaptivity for Finite Elements* (I. Babuška, O.C. Zienkiewicz and E. Arantes e Oliveira, eds.), Wiley, New York, 1985.

[35] P. A. Raviart and J. M. Thomas, "A mixed finite element method for second order elliptic problems," in *Mathematical Aspects of the Finite Element Method*, Springer-Verlag, Heidelberg, 1977.

[36] W. C. Rheinboldt and C. K. Mesztenyi, "On a data structure for adaptive finite element mesh refinements," *TOMS*, **6**(1980), 166–187.

[37] T. F. Russell, *An Incomplete Iterated Characteristic Finite Element Method for a Miscible Displacement Problem*, Ph.D. Thesis, University of Chicago, 1980.

[38] T. F. Russell, "Finite elements with characteristics for two-component incompressible miscible displacement," SPE 10500, *Sixth SPE Symp. on Reservoir Simulation*, New Orleans, 1982.

[39] T. F. Russell and M. F. Wheeler, "Finite element and finite difference methods for continuous flows in porous media," in *Mathematics of Reservoir Simulation* (R. E. Ewing, ed.), SIAM Publications, Philadelphia, 1983, 35–106.

[40] A. Weiser, "Local-mesh, local-order, adaptive finite element methods with a posteriori error estimates for elliptic partial differential equations," *Technical Report No. 213*, Dept. of Computer Science, Yale University, New Haven, Conn., 1981.

[41] M. F. Wheeler and R. Gonzalez, "Mixed finite element methods for petroleum reservoir engineering problems," in *Computing Methods in Applied Science and Engineering 6* (R. Glowinski and J. L. Lions, eds.), North-Holland, Amsterdam, 1984.

Richard E. Ewing, appointed to the University of Wyoming as the J. E. Warren Distinguished Professor of Energy and Environment, holds a joint position in the Departments of Mathematics and *Petroleum Engineering and is Director of the University of Wyoming's Enhanced Oil Recovery Institute. He came to the University from a position as coordinator of mathematical analysis for Mobil Research and Development Corporation in Dallas. After receiving his Ph.D. under the direction of Professor John Cannon at the University of Texas at Austin in 1974, Dr. Ewing held academic positions at Oakland University and the Ohio State University and visiting positions at the University of Chicago and the Mathematics Research Center in Madison, Wisconsin. His principal research interests are in applied mathematics, in numerical analysis and solution of partial differential equations, and in numerical reservoir simulation.*

SYMBOLIC MANIPULATION

HARLEY FLANDERS

Department of Mathematics
The University of Michigan
Ann Arbor, Michigan 48109

Dedicated to Gail Young

1. Introduction.

My object is to explain to mathematicians how a computer can be programmed for a task that is neither number crunching nor data processing. I focus on the particular problem of programming a differentiation machine.

Input to the machine will be any string of characters that can reasonably be interpreted as a real function F(X). For example

$$(A + X)*(5.38 - EXP \ X) + 2*SQR \ SIN(PI*X)$$

Note that functions like EXP and SIN from a preassigned collection of standard functions are allowed, that function arguments may or may not be put inside fences as desired, that functions may be composed, and that standard binary operations may be used freely. Also both symbolic constants like A and PI and real constants may be used.

Output will be a similar string of characters that can reasonably be interpreted as the derivative F'(X) of F(X). Thus we may represent our machine by Figure 1.

The machine will have some way of representing a function F(X) so that the rules of differentiation can be applied. This means that the machine must first translate the input string into its internal representation of a function, a process called *parsing*. Next it must apply the differentiation rules to produce the derivative of the input function as the internal representation of a second function. Finally it must translate this (internal) function into a recognizable string. Thus the differentiation machine naturally decomposes into three submachines, as suggested in Figure 2.

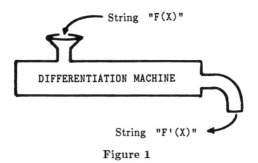

Figure 1

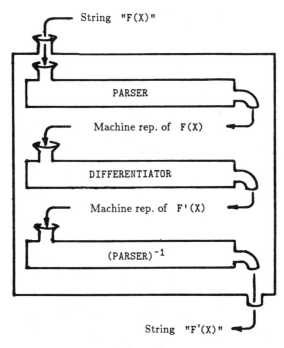

Figure 2

Several issues come up immediately. First, how do we organize a portion of the computer's memory to represent a function, and how do we parse an input string? These issues are discussed in Section 3. Next, precisely what do we mean by a string of characters that can be interpreted as a function? As we shall see later, a precise definition of a function (Section 2) leads almost automatically to a parsing subprogram.

Both differentiation and the inverse of parsing are relatively easy components to program, and we cover them in Sections 4 and 5. My thinking in this subject has been influenced by the stimulating article of Calmut and di Scala [2].

2. What is a Function?

We shall think of a function as an (inverted) tree. Besides the root of the tree, which represents $F(X)$, there are three types of nodes:

(a) Terminal nodes, representing real or symbolic constants or the variable X;

(b) Binop nodes, each with a pair of descending pointers (branches);

(c) Unop nodes, each with a single descending pointer.

An example is shown in Figure 3; it represents the function:

$$F(X) = 3*X + 2*COS(PI*X)$$

We need a very precise definition of function that leaves nothing ambiguous. The definition must include all conceivable parsing options. Thus

$$A + X*B$$

is to be interpreted as

$$A + (X*B), \quad not \quad (A + X)*B,$$

and

$$TAN \; A*X$$

is to be interpreted as

$$TAN(A*X), \quad not \quad (TAN \; A)*X.$$

Our definition of function will be recursive; it is convenient to introduce a metalanguage, *Extended Bachus-Naur Formalism* (EBNF), for expressing this definition. See Wirth [4].

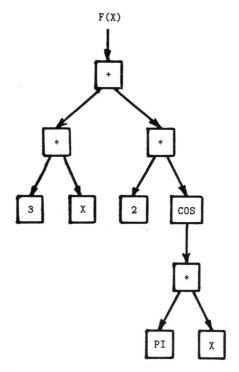

Figure 3

A definition is a finite sequence of statements

```
new object = expression
```

where "=" may be read "is defined as". The expressions consist of a list of primitive expressions (in our case, **constants**, X, SIN, COS, etc.) and other expressions built by certain operations on expressions.

If H and K are expressions, then:

(1) The *concatenation*

```
HK,
```

"H followed by K", is an expression.

(2) The *alternative*

     `H|K,`

"either H   or   K", is an expression.

(3) The *option*

     `[H],`

"either nothing or   H", is an expression.

(4) The *repetition*

     `{H}`

is an expression. It denotes either nothing, or   H, or the concatenation of any number of   H's.

In EBNF, ordinary parentheses ( ) have their usual meaning of grouping, and ordinary double quotes " " are used to enclose an expression to be taken literally.

To get the idea, suppose we have already defined a *term*; think of a term as a product of factors. We then want to define a function as a sum or difference of terms, including at least one summand. Then a corresponding EBNF sentence is

$$\text{function = term \{("+"|"-") term\},}$$

which simply means

$$\text{function = term } \pm \text{ term } \pm \text{ term ...}$$

An equivalent alternative definition is

$$\text{function = term | function ("+"|"-") term,}$$

and yet another is

$$\text{function = [function ("+"|"-")] term.}$$

The latter two, of course, are recursive definitions. Note that these definitions imply left-to-right grouping:

$$\text{H + K + L + M}$$

is grouped

$$\text{((H + K) + L) + M.}$$

We are ready for the definition of a function. Of course "*" means multiplication; also "^" means exponentiation, that is, $5\text{^}3 = 5^3 = 125.$

Definition of a function (* string point of view *)

function	=	term {("+"|"-") term}
term	=	power {("*"|"/") power}
power	=	{factor "^"} factor
factor	=	"X" | constant | "(" function ")" |
		"[" function "]" | "{" function "}" |
		standard_function term | "-" term |
		predefined_function
constant	=	"a" | "b" | "c" | "e" | "pi" | real
standard_function	=	"SIN" | "COS" | "EXP" | "ATAN" | ...
real	=	integer ["." [unsigned_integer]] |
		"." unsigned_integer
predefined_function	=	"f" | "df" | "g" | "dg"

We omit the definitions of unsigned_integer and integer, which the reader can supply as exercises. The predefined_functions are for internal use only; they are generated by the differentiator. Note that powers, if not otherwise grouped, are parsed right-to-left:

$$2\text{^}5\text{^}3 = 2\text{^}(5\text{^}3) = 2^{125},$$

not

$$(2\text{^}5)\text{^}3 = 2\text{^}(5*3) = 2^{15}.$$

If you have not previously worked with such definitions, it will help you to "play computer" with a few expressions and find what they mean. For instance, try

(a) A*B^C/D

(b) A^B/C/D^E

(c) SIN X*COS 5*X

(d) COS X*SIN 5 - X + TAN 2*X

The answers are given at the end of this article.

Note that at least one blank space is understood in concatenations, except where fences are included. We would not write SINX, but rather SIN X, although SIN(X) is acceptable.

3. The Parser.

Suppose the input string is

$$X - COS \; SQRT(1 - X*X)$$

The parser will read this as a sequence of *tokens*

$$\text{"X", "-", "COS", "SQRT", "(", "1", "-", "X", "*", "X", ")".}$$

A friendly parser would do some error checking before doing anything else, checking for unmatched fences, misspellings, etc. We omit discussing this tedious aspect of programming.

Let us assume then that each input string is a sequence of tokens that correctly defines a function. The task of the parser is to construct, from such strings, internal machine representations of functions. We must digress on how the kind of binary tree shown in Figure 3 can be handled in a computer's memory.

We may think of computer memory as a sequence of bytes. (I remind the neophyte that a byte can take $2^8 = 256$ different values, but this is not important here.) High-level programming languages will automatically organize finite sequences of bytes into various data types, like character strings, real numbers, integers, arrays (matrices) of reals, etc. Languages like COBOL, Pascal, and their derivatives allow new data types to be constructed as cartesian products of certain other data types, and the individual chunks of data are called *records*. For instance an indexed complex number can be defined as a record consisting of an index of integer type and a pair of components of real type. That is, the data type is

$$\mathbf{Z} \times \mathbf{R} \times \mathbf{R},$$

and a typical way the record is organized is shown in Figure 4. Each byte in memory is located by its address, an integer whose range depends on the size of the computer. A record like that in Figure 4 is usually referenced by the address of its first byte.

We use records for the nodes of the binary trees representing functions. These records are from the cartesian product of the following eight data types:

KIND	=	{BINOP, UNOP, SYMB_CONS, CONSTANT, X}
BINOP_LIST	=	{PLUS, MINUS, TIMES, DIVIDE, POWER}
UNOP_LIST	=	{SQR, SIN, COS, EXP, etc.}
SYMB_CONS_LIST	=	{A, B, C, E, PI}
CONSTANT_LIST	=	real
LEFT_OPERAND, RIGHT_OPERAND	=	ADDRESS
ARGUMENT	=	ADDRESS

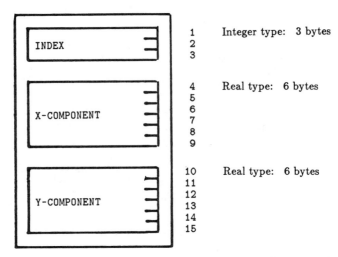

Figure 4. 15 consecutive bytes organized into a record representing an indexed complex number.

The KIND component is essential information for every node. Its value determines which other components carry meaningful information. For instance, if the KIND component is X, then the record represents a terminal node for the "variable" X and no other information is needed. If the KIND component is BINOP, then we must know which of the five possible characters in BINOP_LIST is to apply, and the memory addresses of both its left and its right operands. (For anyone interested, on a modest size computer, each record might require

$$1+1+1+1+6+3+3+3 = 19$$

bytes of memory, which is hardly anything. On a microcomputer with very limited memory, some data compression is worthwhile.)

Let us be specific, and write a corresponding Pascal type declaration. The boldface words are those from the Pascal vocabulary (reserved words). See Flanders [3]. The vertical arrow ↑ is read "pointer to" or "address of". (Note that the declaration is recursive.)

```
type NODE  =  record
                KIND: (BINOP, UNOP, SYM_CON, CON, X);
                B_TYPE: (PLUS, MINUS, etc.);
                U_TYPE: (SQR, SIN, etc.);
                SYM: (A, B, C, E, PI);
                COEFF: Real;
                LEFT, RIGHT, ARG: ↑NODE
            end;
```

A quick explanation: in Pascal, the notation

$$(A, B, C, E, PI)$$

defines a *scalar* (enumerated) type whose members are the five *indentifiers* A,...,PI.

Now we can define type FUNCTION to be a pointer type, a pointer to a NODE. Actually we can't use the word FUNCTION because **function** is a Pascal reserved word. Instead

type FUNCTION_OF_X = ↑NODE;

Suppose we declare F to be a variable of type function; i.e., a function as we understand it:

var F: FUNCTION_OF_X;

Suppose in the course of program execution, F is assigned—given a value. Then F↑ denotes the "pointee" of F, the node that F is the address of. The KIND component

F↑.KIND

of F↑ has one of the five possible values BINOP,...,X. Suppose

F↑.KIND = UNOP.

Then

F↑.U_TYPE

has one of the values SQR,SIN,..., say

F↑.U_TYPE = TAN.

Finally F↑.ARG is the address of another node, that is, it is another function G(X) and

$$F(X) = (TAN \circ G)(X).$$

There are several popular methods of constructing parsers. For a lengthy discussion of the theory see Aho and Ullman [1]. We shall base a parser on what is called *recursive descent*. We now describe part of the program that parses input strings, and closely follow the EBNF formal definition of a function in Section 2. The (Pascal) program is a sequence of nested procedures_with_output, in Pascal called *functions*. Each has the same kind of input, a string and an index of where we are in the string. The output of each function is a FUNCTION_OF_X, and the nesting is

function EXPRESSION(<input>): <output>;

function TERM(<input>): <output>;

function POWER(<input>): <output>;

function FACTOR(<input>): <output>;

I use EXPRESSION instead of "function" to avoid overloading that word. Everything in sight is recursive, and the innermost procedure FACTOR will call the outermost EXPRESSION if the next token happens to be one of the opening fences (, [, or {.

Suppose we have finished programming up to the level TERM. When we call EXPRESSION, we are dealing with the input string and the index of the first character of the next token in the string. For instance, suppose the input string is

S = '3*X + 2*COS(PI*X)'

(* 12345678901234567 *)

(The string quotes ' ' are not part of the string.) This is a string of 17 characters, including two spaces. The integer variable INDEX varies from 1 to 17. When INDEX = 1, EXPRESSION is invoked, and again when INDEX = 13 because an opening parenthesis has just been passed. Thus our procedure declaration must be

function EXPRESSION

(S: string; **var** INDEX: integer): FUNCTION_OF_X;

Each execution of the procedure EXPRESSION, in addition to producing an output, must increase the input variable INDEX to the beginning of the next token in S after the execution. That accounts for the reserved word **var**.

The procedure EXPRESSION needs some stuff to work with, a couple of temporary functions_of_X, say G and H, which will be declared as variables of type FUNCTION_OF_X. Pascal contains a standard procedure, New, which initializes any pointer variable in the sense that it assigns an adequate chunk of memory for the pointee of that variable, and assigns its address. Let us write the declaration:

```
function EXPRESSION(S: string; var INDEX: integer):  FUNCTION_OF_X;
    var     (* Comment: variables local to EXPRESSION *)
        G, H: FUNCTION_OF_X;
        TOKEN: string;
    function TERM ...;
        (* Comment: we assume this program written *)
    begin     (* Comment: the action of EXPRESSION *)
    G := TERM(S, INDEX);
    TOKEN := NEXT_TOKEN;
        (* Comment: we assume NEXT_TOKEN is a procedure whose output
            is the next token in S, starting at INDEX. Its execution must
            increase INDEX to the start of the following token. *)
    while (TOKEN = '+') or (TOKEN = '-') do
        begin     (* while loop *)
        H := G; New(G); G↑.KIND := BINOP;
        if TOKEN = '+' then G↑.B_TYPE := PLUS
        else G↑.B_TYPE := MINUS;
            (* Point P; see Figure 5 *)
        G↑.LEFT := H; G↑.RIGHT := TERM(S, INDEX)
        TOKEN := NEXT_TOKEN
        end;     (* while loop *)
    EXPRESSION := G
    end;     (* EXPRESSION *)
```

For example, suppose

$$S = \text{'a - TAN X + b'}$$
$$(* \ 1234567890123 \ *)$$

We assign INDEX := 1 and call

$$F = \text{EXPRESSION(S, INDEX)}.$$

We show in Figure 5 various steps in the execution of this call.

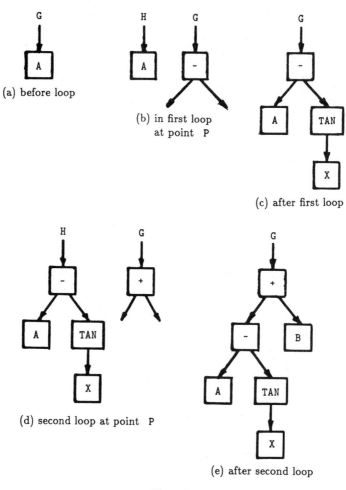

(a) before loop

(b) in first loop
at point P

(c) after first loop

(d) second loop at point P

(e) after second loop

Figure 5

Note that in general, EXPRESSION builds the kind of tree shown in Figure 6.

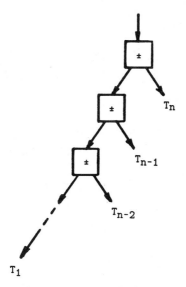

Figure 6. The T_i are terms.

Note how the action of EXPRESSION closely follows the EBNF definition for function, and how really routine a piece of program it is. Programs for TERM and POWER are similar. The program for FACTOR is, of course, longer because it covers many cases, and must deal with terminals.

4. The Differentiator.

First we must write down all of the rules. This takes the form of an array DER of strings, indexed by the various unops and binops. Here are two typical entries:

DER[TIMES] = 'df*g+f*dg'

DER[TAN] = 'df*SQR SEC f'

The program for derivative has the declaration

function DERIVATIVE(F: FUNCTION_OF_X): FUNCTION_OF_X;

so its input is a pointer to a node, as is its output. Its action is recursive; it calls itself.

Suppose for instance that the input H has the form

$$F(X)*G(X)$$

that is, it is a pointer to a TIMES node, as shown in Figure 7a. The action of the differentiator will be the following steps:

1. Assign S := DER[TIMES]; INDEX := 1;

2. (Recursively) assign DF := DERIVATIVE(F);
   $\qquad\qquad\qquad\qquad$ DG := DERIVATIVE(G);

3. Assign DERIVATIVE := EXPRESSION(S, INDEX);

Thus the string S is sent through the parser! I postponed until now noting that part of the action of FACTOR is to replace the tokens f, g, df, dg by the pointers F, G, DF, and DG when it builds a tree.

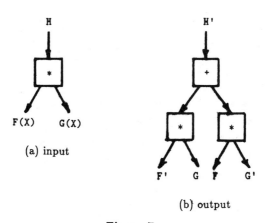

(a) input

(b) output

Figure 7

For a second example, showing the chain rule, consider

$$H(X) = TAN\ F(X)$$

The steps are

1. S := DER[TAN] = 'df*SQR SEC f'; INDEX := 1;

2. DF := DERIVATIVE(F);

3. DERIVATIVE := EXPRESSION(S, INDEX);

See Figure 8.

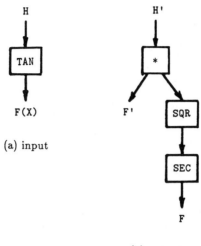

(a) input

(b) output

Figure 8

5. The Inverse_Parser.

We started with a character string S. The parser built a binary tree F and the differentiator built a second binary tree F'. Now we want to change F' into a string T representing the same function F', so we can print it.

One issue I have ignored is simplification. That really is a large topic that would require another talk, indeed, it is more complicated and dreadfully more tedious than this present story. However, to indicate what we have to face, consider the input string

$$S = \text{'SQR SIN X + SQR COS X'}.$$

Suppose we do not simplify this before parsing and differentiating. The resulting binary tree will most likely inverse_parse to

$$T = \text{'2*(COS X)*SIN X + 2*(-SIN X)*COS X'},$$

surely an unsatisfactory answer. Enough said.

In parsing, our object is a string, and we traverse that string character-by-character, left-to-right. The definition of function in Section 2 was sensitive to this input and the way we scan it.

For `inverse_parsing` we have a very different situation. Our input is a pointer, the root of a tree. It points to the highest node in the tree, and that is either a terminal, which we just want to print, or it is a unop, whose name we want to print and then we want to `inverse_parse` its argument, or it is a binop. In that case our action is (1) `inverse_parse` its left argument, (2) print the binop's name, then (3) `inverse_parse` its right argument.

Our definition of function in Section 2 was motivated by the need to parse strings. This present discussion of how we want an `inverse_parser` to act suggests another definition of function, one fitted to the tree representation, and strongly influenced by the way we access a tree: through a node.

Definition of a function (* tree point of view *)

```
function            =   constant | "X" | standard_function function |
                        function ("+" | "-" | "*" | "/" | "^") function

standard_function   =   ... etc.

constant            =   "a" | "b" | "c" | "e" | "pi" | real

real                =   ... etc.
```

It is now fairly routine to write a procedure with input a tree and output a string. The only real problem is when to use parentheses. (It would be pretty hard to use parentheses, brackets, and braces in a sensible fashion and I have never tried to do so, nor have I ever seen a program that does.) The tree in Figure 9a does not require fences; that in Figure 9b obviously does. So deciding when to use fences requires some look-ahead, and that increases program length considerably. Another output problem that must be faced in the real world is where to insert line breaks. We shall ignore this problem here, and to simplify programming, insert parentheses everywhere, needed or not. Then the `inverse_parser` program looks something like this:

```
procedure PRINT(F: FUNCTION_OF_X);
    begin
    case F↑.KIND of
        BINOP:
            begin
            Write('('); Print(F^.LEFT); Write(')');
            case F^.B_TYPE of
                PLUS: Write('+');
                etc.
                end;    (* case *)
```

```
        Write('('); Print(F^.RIGHT); Write(')')
    end;    (* BINOP *)
UNOP: etc.

...

X: Write('X')
end    (* cases of F^.KIND *)
end;    (* PRINT *)
```

Note that the procedure is recursive, and it follows the definition of function. Note also that we have the omitted the string, but gone directly to the output file, which avoids a few technicalities of string manipulation. Also we have used the standard Pascal procedure Write for character output. In the UNOP case it is also used to write strings, 'SIN', etc.

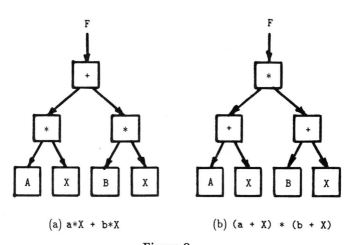

(a) a*X + b*X   (b) (a + X) * (b + X)

Figure 9

Finally, I show in Figure 10 an example of input and output from a complete program, including that dreadful simplification.

$$F(x) = X*SINH(A*X)$$

$$dF/dx = SINH(A*X)+A*X*COSH(A*X)$$

$$d2F/dx2 = 2*A*COSH(A*X)+(A^2)*X*SINH(A*X)$$

$$d3F/dx3 = 3*(A^2)*SINH(A*X)+(A^3)*X*COSH(A*X)$$

Figure 10. The proof of the pudding.

Answers to the Exercises of Section 2.

(a) $\dfrac{AB^C}{D}$ 　　　　 (b) $\dfrac{A^B}{CD^E}$

(c) 　　 SIN[X(COS 5X)]

(d) 　　 COS(X·SIN 5) - X + TAN 2X

References.

[1] Alfred V. Aho and Jeffrey D. Ullman, *Principles of Compiler Design*, Addison-Wesley Pub. Co., 1977 (especially Chapters 5 and 6).

[2] J. Calmet and R-M. di Scala, "Pascal as host language of computer algebra systems," *SIGPLAN Notices*, **18** (July 1983), 15–24.

[3] Harley Flanders, *Scientific Pascal*, Reston Publishing Co. (Prentice-Hall), 1984.

[4] Niklaus Wirth, *Programming in Modula-2*, 2nd ed., Springer-Verlag, 1983 (especially pp. 10–11).

The author's University of Chicago dissertation, under A. Weil and O. F. G. Schilling, was on class field theory. After a Bateman Fellowship at Cal Tech he spent some years at Berkeley, moving into differential geometry. At Purdue he continued work in algebra and geometry and contributed to electric circuit theory. This work continued at Tel Aviv and Florida Atlantic universities, where computing needs naturally pushed him into scientific computation. He served as Editor of the American Mathematical Monthly and has authored several text books.

SOME UNIFYING CONCEPTS IN APPLIED MATHEMATICS

ISMAEL HERRERA

Instituto de Geofisica
Universidad Nacional de Mexico
Apartado Postal 21-524
04000 Mexico, D.F.

1. Introduction.

I was trained, originally, as a pure mathematician although I have always worked in applications. Since I was a student I have recognized that the methodology of mathematical thinking is very powerful as a tool for development. It has, indeed, great practical value. Henry Pollak in his excellent talk today presented many interesting examples which seem to indicate that in the field of communications this is thoroughly recognized. However, there are still many areas of industry and of other human endeavors in which skepticism is prevalent.

Those who are in doubt about the practical value of clarity should be asked to work in darkness; hence, the usefulness of rigor which is the pathway to clarity. When clarity is granted, it is easy to achieve conceptual unity since the basic properties on which other properties depend then become apparent and the formulation of assertions of general validity becomes feasible. The practical value of generality is clear, because of the economy of effort that is achieved through the formulation of statements of general validity.

But for my taste, the most highly appreciated of all the characteristics of mathematical thinking is simplicity. The capacity for transforming complicated things into simple ones is a power which seems to spring from a magician. And simplicity has a tremendous practical value, because when concepts become simple our capacity to handle them grows.

It is with this conviction that I have applied mathematical methodology to the very applied questions I have faced in my professional practice. It was also in this spirit that I developed the exercise whose results I am going to present today.

More than ten years ago, when I got back to the National University of Mexico after having organized the National Council for Science and Technology in my country, I started work in numerical methods. I was captivated by the growing power of mathematical modeling which is drastically transforming science and engineering due to its capacity to mimic macroscopic systems. This led me to develop some unifying concepts of numerical methods for partial differential equations. Some of the practical implications have been reported in a sequence of papers [1,2].

Here, I will restrict my attention to the abstract formulation of this algebraic theory. I include some results which are more general than those I had reported previously. In particular, I present the theory for operators $P : D_1 \to D_2^*$, while previously I had restricted attention to operators $P : D \to D^*$. Also, Theorem 4.1 is new and will simplify the application of the theory to finite element methods.

2. Boundary Operators and Formal Adjoints.

Let D_1 and D_2 be two linear spaces and write D_1^* and D_2^* for the corresponding algebraic duals. Consider a linear operator $P : D_1 \to D_2^*$, then for every $u \in D_1$ and $v \in D_2$ one has that $Pu \in D_2^*$ while $\langle Pu, v \rangle \in \mathcal{F}$, where $\mathcal{F}$ is the field of scalars. For applications to partial differential equations this will be either the real or complex numbers. With the transposed bilinear form $\langle Pv, u \rangle$ we associate the transposed operator $P^* : D_2 \to D_1^*$. Hence

$$\langle P^*u, v \rangle = \langle Pv, u \rangle \qquad (2.1)$$

Generally, we consider two classes of operators $P : D_1 \to D_2^*$ and $Q : D_2 \to D_1^*$; when two operators are in the same class, we say that they are of the same kind; otherwise they are of opposite class.

The notion of boundary operator, in this abstract setting, is introduced as a relation among such operators. Thus, given two operators $B : D_1 \to D_2^*$ and $P : D_1 \to D_2^*$, we say that B is a boundary operator for P, when for every $u \in D_1$,

$$\langle Pu, v \rangle = 0 \qquad \forall v \in N_{B^*} \quad \implies \quad Pu = 0 \qquad (2.2)$$

Here, N_{B^*} stands for the null subspace of B^*.

To illustrate the kind of situations we have in mind, consider a region Ω (Figure 1) and a linear differential operator $\mathcal{L}$ defined in a suitable linear space of functions (e.g., $H^s(\Omega)$ with s sufficiently large). Define

$$\langle Pu, v \rangle = \int_\Omega v \mathcal{L}u \, dx; \qquad \langle Bu, v \rangle = \int_{\partial\Omega} v \frac{\partial u}{\partial n} \, dx \qquad (2.3)$$

Then, it is easy to verify that for usual choices of the spaces D_1 and D_2, one has $N_{B^*} = \{v \in D_2 \mid v = 0 \text{ on } \partial\Omega\}$. Using this result it is clear that B is a boundary operator for P (indeed, this follows from the so-called fundamental theorem of calculus of variations).

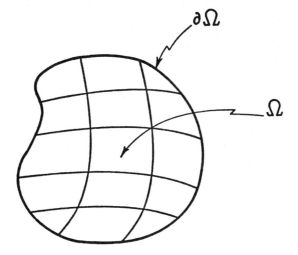

Figure 1

Once the notion of boundary operator has been introduced, one can define formal adjoints. Indeed, let P and Q be of opposite kind, we say that P and Q are formal adjoints when $R = P - Q^*$ is a boundary operator for P while R^* is a boundary operator for Q. This notion introduces a symmetric relation between operators of opposite kinds.

To illustrate this notion, consider a differential operator $\mathcal{L}$ and its formal adjoint $\mathcal{L}^*$, in the usual sense. In addition to P (Eq. 2.3), consider $Q : D_2 \to D_1^*$ given by

$$\langle Qu, v \rangle = \int_\Omega v \mathcal{L}^* u \, dx \qquad (2.4)$$

Then

$$\langle (P - Q^*)u, v \rangle = \langle Ru, v \rangle = \int_{\partial\Omega} \text{boundary terms } dx \qquad (2.5)$$

Again, for usual choices of D_1 and D_2, one has that P and Q are formal adjoints.

3. Green's Formulas.

Once the notion of formal adjoints is available, one can introduce Green's formulas. For differential operators a relation of the form (see Lions and Magenes [3]):

$$\int_\Omega v \mathcal{L} u \, dx - \int_{\partial\Omega} B(u, v) \, dx = \int_\Omega u \mathcal{L}^* v \, dx - \int_{\partial\Omega} C^*(u, v) \, dx \qquad (3.1)$$

is said to be a Green's formula. Thus, it would be natural to define Green's formula, in our setting, by

$$P - B = Q^* - C^* \qquad (3.2)$$

i.e., $R = B - C^*$, when B is boundary operator for P, while C is boundary operator for Q. However, such definition would be too loose and not very useful. Just as in the case of differential operators [Lions and Magenes, 1972], we will require that the decomposition $R = B - C^*$ satisfy additional conditions which are introduced next.

Given two operators R_1 and R_2 of the same kind, we say that R_1 and R_2 are disjoint when R_2 is boundary operator for R_1 while R_1 is boundary operator for R_2. In addition, R_1 and R_2 are fully disjoint when

a) R_1 and R_2 are disjoint;

b) R_1^* and R_2^* are disjoint.

Finally, let R_1 and R_2 be fully disjoint, while

$$R = R_1 + R_2 \qquad (3.3)$$

Then we say that the pair $\{R_1, R_2\}$ decomposes (weakly) R.

Example. Take $D_1 = D_2 = C[0,2]$. Define

$$\langle Ru, v \rangle = \int_0^2 uv\, dx$$

$$\langle R_1 u, v \rangle = \int_0^1 uv\, dx; \qquad \langle R_2 u, v \rangle = \int_1^2 uv\, dx$$

Then R_1 and R_2 are fully disjoint. Hence, $\{R_1, R_2\}$ decomposes R, because (3.3) is obviously satisfied.

Remark. When (3.3) holds, it is clear that $N_R \supset N_{R_1} \cap N_{R_2}$. However, when $\{R_1, R_2\}$ decomposes R one has in addition that

$$N_R = N_{R_1} \cap N_{R_2} \tag{3.4}$$

We are now ready to define Green's formulas. Let P and Q be formal adjoints, then the relation (3.2) is said to be a Green's formula when B and C^* are fully disjoint. Clearly, in such a case the pair $\{B, C^*\}$ is a weak decomposition of $R = P - Q^*$. Also, B is a boundary operator for P, while C is a boundary operator for Q.

4. Abstract Green's Formulas.

In this section an abstract characterization of Green's formulas [2], which is quite useful in applications of the theory, is presented.

Remark 4.1. Assume R_1 and R_2 decompose weakly R. Then, it is easy to verify that

a) $\langle Ru, v \rangle = 0 \qquad \forall u \in N_{R_1}$ and $v \in N_{R_2^*}$ $\tag{4.1}$

b) $N_{R_1} \supset N_R$ and $N_{R_2^*} \supset N_{R^*}$ $\tag{4.2}$

c) $\langle Ru, v \rangle = 0 \qquad \forall v \in N_{R_2^*} \qquad \Longleftrightarrow \qquad u \in N_{R_1}$ $\tag{4.3a}$

and

$\langle Ru, v \rangle = 0 \qquad \forall u \in N_{R_1} \qquad \Longleftrightarrow \qquad v \in N_{R_2^*}$ $\tag{4.3b}$

Thus, given linear subspaces $I_1 \subset D_1$ and $I_2 \subset D_2$, we say that the pair $\{I_1, I_2\}$ is:

i) Conjugate when

$$\langle Ru, v \rangle = 0 \qquad \forall u \in I_1 \quad \text{and} \quad v \in I_2 \tag{4.4}$$

ii) Regular conjugate when

$$I_1 \supset N_R \qquad \text{and} \qquad I_2 \supset N_{R^*} \tag{4.5}$$

in addition to (4.4).

iii) Completely regular when

$$\langle Ru, v \rangle = 0 \qquad \forall v \in I_2 \qquad \Longleftrightarrow \qquad u \in I_1 \qquad (4.6a)$$

and simultaneously

$$\langle Ru, v \rangle = 0 \qquad \forall u \in I_1 \qquad \Longleftrightarrow \qquad v \in I_2 \qquad (4.6b)$$

Remark 4.2. Every completely regular pair is a regular conjugate pair.

Applying these definitions, it is seen that each of the pairs $\{N_{R_1}, N_{R_2^*}\}$ and $\{N_{R_2}, N_{R_1^*}\}$ are regular conjugate pairs. Even more, they are completely regular. Observe that

$$N_{R^*} = N_{R_1^*} \cap N_{R_2^*} \qquad (4.7)$$

in addition to (3.4).

In view of the above, we give the following definition of abstract weak decomposition.

Definition 4.1. Let $R : D_1 \to D_2^*$ be an operator. Then two ordered pairs $\{I_{11}, I_{22}\}$ and $\{I_{12}, I_{21}\}$ of completely regular conjugate subspaces is said to be an abstract weak decomposition of R when

$$N_R = I_{11} \cap I_{12} \qquad and \qquad N_{R^*} = I_{21} \cap I_{22} \qquad (4.8)$$

Remark 4.3. Given a weak decomposition $\{R_1, R_2\}$ of R, define

$$I_{11} = N_{R_2}; \qquad I_{22} = N_{R_1^*} \qquad (4.9a)$$

$$I_{12} = N_{R_1}; \qquad I_{21} = N_{R_2^*} \qquad (4.9b)$$

Then $\{I_{11}, I_{22}\}, \{I_{12}, I_{21}\}$ is an abstract weak decomposition of R.

Generally, given an abstract weak decomposition $\{I_{11}, I_{22}\}, \{I_{12}, I_{21}\}$ of R, we say that a pair of operators $\{R_1, R_2\}$ ($R_1 : D_1 \to D_2^*$ and $R_2 : D_1 \to D_2^*$) is a realization of the abstract decomposition when equations (3.3) and (4.9) are satisfied.

Remark 4.4. It is easy to verify that every realization $\{R_1, R_2\}$ of an abstract weak decomposition is a weak decomposition of R.

Theorem 4.1. Assume $\{I_{11}, I_{22}\}, \{I_{12}, I_{21}\}$ is an abstract weak decomposition of R. Then:

i) There exist realizations of the abstract w-decomposition.

ii) A pair of operators $\{R_1, R_2\}$ such that $R = R_1 + R_2$ is a realization of the weak decomposition, if and only if

$$I_{11} \subset N_{R_2}; \qquad I_{22} \subset N_{R_1^*} \qquad (4.10a)$$

$$I_{12} \subset N_{R_1}; \qquad I_{21} \subset N_{R_2^*} \qquad (4.10b)$$

iii) Let $D_1^0 = I_{11} + I_{12}$, $D_2^0 = I_{21} + I_{22}$, and take $D' \subset D_1$, $D'' \subset D_2$ as algebraic complements of D_1^0 and D_2^0, respectively. Then $\{R_1, R_2\}$ is a realization of the abstract decomposition, if and only if

$$\langle R_1 u, v \rangle = \langle R(u - u_{12}), v - v_{22} \rangle + \langle \delta_1 u, v \rangle \tag{4.11a}$$

and

$$\langle R_2 u, v \rangle = \langle R(u - u_{22}), v - v_{21} \rangle + \langle \delta_2 u, v \rangle \tag{4.11b}$$

where δ_1, δ_2 are any two operators with the property

$$u \in D_1^0 \quad or \quad v \in D_2^0 \quad \Longrightarrow \quad \langle \delta_1 u, v \rangle = \langle \delta_2 u, v \rangle = 0 \tag{4.12}$$

and such that $\delta_1 + \delta_2 = \delta$. Here, $\delta : D_1 \to D_2^*$ is defined by

$$\langle \delta u, v \rangle = -\langle Ru', v'' \rangle. \tag{4.13}$$

Proof. The proof of this result is given in [4].

Let P and Q be formal adjoints and $R = P - Q^*$. Then, every abstract weak decomposition of R is said to be an abstract Green's formula. Associated with every abstract Green's formula $\{I_{11}, I_{22}\}$, $\{I_{12}, I_{21}\}$ there exist Green's formulas $P - B = Q^* - C^*$ such that

$$I_{11} = N_{C^*}; \qquad I_{22} = N_{B^*} \tag{4.14a}$$

$$I_{12} = N_B \; ; \qquad I_{21} = N_C. \tag{4.14b}$$

Any such Green's formula is given by

$$\langle Bu, v \rangle = \langle R(u - u_{12}), v - v_{22} \rangle + \langle \delta_1 u, v \rangle \tag{4.15a}$$

$$\langle Cu, v \rangle = -\langle R(v - v_{11}), u - u_{21} \rangle + \langle \delta_2 u, v \rangle. \tag{4.15b}$$

5. Strong Green's Formulas.

In applications, Green's formulas are more easily constructed with the aid of a related concept here explained.

Definition 5.1. Two operators R_1 and R_2 of the same kind (for definiteness, they are here assumed to be $D_1 \to D_2^*$) are said to have independent variations when, for every $U \in D_1$ and $V \in D_1$, there exists $u \in D_1$ such that

$$R_1 u = R_1 U \qquad and \qquad R_2 u \in R_2 V \tag{5.1}$$

They have fully independent variations when, in addition, R_1^* and R_2^* also have independent variations.

It is easy to see that the operators $R_1 : D_1 \to D_2^*$ and $R_2 : D_1 \to D_2^*$ have independent variations, if and only if [2]

$$D_1 = N_{R_1} + N_{R_2} \tag{5.2a}$$

They have fully independent variations when in addition

$$D_2 = N_{R_1^*} + N_{R_2^*} \tag{5.2b}$$

Remark 5.1. An important property is that R_1^* and R_2^* are disjoint whenever R_1 and R_2 have independent variations. This implies that R_1 and R_2 are fully disjoint whenever R_1 and R_2 have fully independent variations.

Definition 5.2. *Given operators R, R_1 and R_2 of the same kind such that*

a) R_1 and R_2 have fully independent variations;

b)

$$R = R_1 + R_2 \tag{5.3}$$

one says that the pair $\{R_1, R_2\}$ is an S-decomposition of R (or strong decomposition of R).

It is easy to give an abstract characterization of S-decompositions.

Definition 5.3. *Two regular conjugate pairs $\{I_{11}, I_{22}\}$ and $\{I_{12}, I_{21}\}$ are said to be an abstract S-decomposition of R, when*

$$D_1 = I_{11} + I_{12} \quad \text{and} \quad D_2 = I_{21} + I_{22} \tag{5.4}$$

The basic result is given next.

Theorem 5.1. *There is a one-to-one correspondence between S-decompositions of R and abstract S-decompositions of R. Given an S-decomposition $\{R_1, R_2\}$ of R, the corresponding abstract S-decomposition of R is $\{I_{11}, I_{22}\}$, $\{I_{12}, I_{21}\}$, defined by equations (4.9). Conversely, when $\{I_{11}, I_{22}\}$, $\{I_{12}, I_{21}\}$ is an abstract S-decomposition, the corresponding S-decomposition is the pair $\{R_1, R_2\}$ defined by*

$$\langle R_1 u, v \rangle = \langle R u_{11}, v_{12} \rangle \qquad \langle R_2 u, v \rangle = \langle R u_{12}, v_{22} \rangle \tag{5.5}$$

Proof. This follows easily from the following result.

Lemma 5.1. *Every abstract S-decomposition of R is an abstract w-decomposition of R.*

Proof. When $\{I_{11}, I_{22}\}$, $\{I_{12}, I_{21}\}$ is an abstract S-decomposition of R, one has

i)

$$N_R = I_{11} \cap I_{12} \qquad \text{and} \qquad N_{R^*} = I_{21} \cap I_{22} \tag{5.6}$$

and

ii) both $\{I_{11}, I_{22}\}$ and $\{I_{12}, I_{21}\}$ are completely regular pairs of conjugate subspaces.

This can be seen by a slight extension of the arguments presented in [2]. Then, the Lemma follows by a straightforward application of Definition 4.1. Because of Lemma 5.1, Theorem 4.1 can be applied with $D' = D'' = \{0\}$. Then, Theorem 5.1 is clear by virtue of Remark 4.3 and Theorem 4.1.

Definition 5.2. Let $P : D_1 \to D_2^*$ and $Q : D_2 \to D_1^*$ be formal adjoints. The relation

$$P - B = Q^* - C^* \tag{5.7}$$

is said to be an **S-Green's formula** when the pair $\{B, -C^*\}$ is an S-decomposition of $R = P - Q^*$. Correspondingly, a couple $\{I_{11}, I_{22}\}$, $\{I_{12}, I_{21}\}$ of subspaces is an abstract S-Green's formula when such couple is an abstract S-decomposition of $R = P - Q^*$.

Theorem 5.2. Given P and Q, formal adjoints, there is a one-to-one correspondence between S-Green's formulas and abstract S-Green's formulas. When (5.7) is a Green's formula, the corresponding abstract S-Green's formula is defined by (4.13). Conversely, given the abstract S-Green's formula $\{I_{11}, I_{22}\}$, $\{I_{12}, I_{21}\}$, the corresponding S-Green's formula follows from

$$\langle Bu, v \rangle = \langle Ru_{11}, v_{21} \rangle; \qquad \langle Cu, v \rangle = -\langle Rv_{12}, u_{22} \rangle \tag{5.8}$$

Proof. It follows from Theorem 5.1.

Remark on Notation. When (5.7) is a Green's formula, the abstract Green's formula, defined by (4.14), is said to be its abstraction. Conversely, given an abstract Green's formula, the Green's formula defined by (4.14) is a realization of the abstract Green's formula. Clearly, every Green's formula has a unique abstraction. However, in general abstract Green's formulas have many realizations, except when they are abstract S-Green's formulas, in which case the realization is unique. In this case, the correspondence is one-to-one.

References.

[1] I. Herrera, *Boundary Methods: An Algebraic Theory*, Pitman Advanced Publishing Program, 1984.
[2] I. Herrera, "Unified approach to numerical methods," *Numerical Methods for Partial Differential Equations*, **1**(1985), 25–44 (part I); 159–186 (part II); 241–258 (part III).

[3] J. L. Lions and E. Magenes, *Non-Homogeneous Boundary Value Problems and Applications*, Vol. 1, Springer-Verlag, 1972.

[4] I. Herrera, *Algebraic Theory of Boundary Value Problems, Applicable Analysis*, to be published, 1986.

Dr. Ismael Herrera is Director of the Instituto de Geofisica at the Universidad Nacional Autonoma de Mexico (U.N.A.M.) in Mexico City. He earned a B.Sc. in mathematics and physics at U.N.A.M. and a Ph.D. in applied mathematics at Brown University. He is the author of over 90 scientific papers in mathematics, geophysics, and wave propagation and has written several books. His most recent monograph is Boundary Methods: An Algebraic Theory (Pitman, 1984).

TEACHING AND RESEARCH:
THE HISTORY OF A PSEUDOCONFLICT

PETER HILTON

Department of Mathematical Sciences
State University of New York
Binghamton, New York 13901

To my friend Gail Young,
in affection and admiration

1. Introduction.

I have chosen this topic for its appropriateness both to the theme of the conference—new directions in applied and computational mathematics—and to the occasion on which we honor Gail Young. For the new directions, with which other speakers will be dealing more directly in their contributions, carry strong implications for the teaching of mathematics at graduate and undergraduate levels—indeed, at all levels—so that it is fitting that we pay attention now to those implications. The organizers have, of course, taken note of this aspect of the conference theme, in placing on our agenda, as its final item, a panel discussion on the general issue of the shape of a new mathematics curriculum. I will confine myself to one aspect of the implications for us as teachers of these new directions; and I promise that my remarks today will be neither repeated nor contradicted when I reappear before you as a panelist tomorrow.

The aspect on which I will concentrate is indicated by the title of my address. For I am claiming that the new directions in mathematics itself and in the uses to which mathematics is being put should, profoundly, affect what we teach and the way we teach, should profoundly influence both the content and the style of our teaching. If this proposition—which seems to me self-evidently true—is accepted, then it surely follows that teachers of mathematics should be very much aware of what is happening in mathematics today. I would further argue that such an awareness is far more likely to reflect an active interest, rather than a passive cognizance, of those new developments; and that the teaching, at whatever level, will be the more vivid, exciting and inspiring when the interest of the teacher is active. Thus I am led to argue in favor of the natural coexistence of teaching and research, and to dispute the often-repeated claim that they are in conflict. And thus, too, I find myself choosing this crucial time in the history of our subject to trace the effects of the fallacious view that there is such a conflict. For if those who would separate teaching and research—and, as I will show, such people exercise a powerful, if strident and crude, voice in our academic affairs—if these people are successful in peddling their ill-conceived advice, the effect would be disastrous.

On the other hand, I must admit to a certain grandiosity in my title. I am not a historian, and there will only be some recent history in my remarks. Perhaps I should have entitled my talk "The Pseudo-History of a Conflict"; this, however, would have been to err much more in the direction of honesty than is customary in discursive talks of this nature.

I also argue that Gail Young's long and highly productive professional life attests to the truth of my central proposition—the complementarity of teaching and research. Nobody better epitomizes the harmonious interaction of good teaching and good research as twin aspects of good academic citizenship.

2. An Unfortunate Split.

The (apparent) conflict between teaching and research is one of the classical false dichotomies of our profession, some of which extend far beyond mathematics to embrace virtually every academic discipline. Let us, however, narrow the scope of our discussion to mathematics and mathematics education. In an address at the Karlsruhe meeting of ICME [10], I identified several of these "pseudo-antitheses"; thus

art	*vs*	science
pure mathematics	*vs*	applied mathematics
elitism	*vs*	egalitarianism
problem solving	*vs*	structure-building
classical	*vs*	modern
teaching	*vs*	research

I did not argue then, and do not argue now, that there is no conflict among the practitioners of mathematics with respect to these dichotomies. Self-styled pure and applied mathematicians are sometimes in dispute; those who advocate an emphasis on problem-solving in the pre-college curriculum do often stigmatize those (e.g. the supporters of the "new math") who place emphasis on the study of mathematical structures; those who favor equality of educational opportunity for all our citizens do criticize those who seek to develop programs appropriate to the gifted and talented. My claim is that the conflict is diversionary and that the community of interest between the advocates of the two "sides" should be the dominant influence.

Nowhere, I claim, is all this truer than with respect to the sterile dispute between teaching and research. There is, as a matter of sociological fact, a conflict between teachers and researchers. There is a most unfortunate separation, within our profession, between those identified as predominantly teachers and those recognized as research mathematicians. There is often, and most unfortunately, a sense of resentment within the membership of a given mathematics department of the privileges, prestige and advantages of the more prominent researchers among them; and this is frequently matched by a certain hauteur on the part of these researchers. Yes, conflicts do exist, but I believe them to be misguided, unnecessary and avoidable.

Apartheid is most objectionable, and most difficult to eradicate, when it is institutionalized. It would take a historian far, far more subtle than I to disentangle cause and effect in what I argue to be the single most damaging event in the evolution of this conflict among mathematicians in the U.S.A.—the creation of two separate organizations of professional mathematicians, the A.M.S. and the M.A.A. I believe that, by this fissiparous act, the mathematicians of this country took a giant step backwards and handed to us a legacy of prejudice and erroneous thinking which is plaguing us strongly today.

Of course, I cannot argue simply that the false views today on the need to separate teaching and research were caused by that act in 1917 when the M.A.A. was formed—for what caused the act? It would be as well to sigh nostagically for the state of the world before the First World War—and fail to understand that the state of the world then gave rise to the First World War! But the existence of these two institutions side by side gives permanence, and an appearance of respectability, to the view that research and teaching are separate activities and that each individual mathematician may be identified as representative of one of those disjoint aspects of the profession. There are outstanding examples of mathematicians who bestride the two "cultures"—one thinks of past presidents of the M.A.A. like Henry Pollak, Dick Anderson, Gail Young himself—but these are exceptions to the general rule that most mathematicians are clearly identified with one or the other institution, and thus with a predominant concern with research or teaching.

One does not find parallels for this institutionalized division of our community in many other countries. In Canada, the Canadian Mathematical Society is responsible for fostering both research and good teaching at the undergraduate level; in the U.K. the London Mathematical Society also has this dual responsibility, though admittedly it is identified in the eyes of most mathematicians as a sponsor of research. But there is no separate organization to encourage undergraduate teaching—

the Mathematical Association attracts secondary teachers as its members, although leading mathematical luminaries will often be found in the largely ceremonial role of president.

It may well be argued that the situation is different in the U.S.A. since here, unlike Canada and the U.K., we have a system of tertiary education comprising both universities and colleges, that is, institutions with graduate programs and purely undergraduate institutions. It may indeed be true that this two-tier system may seem to some—and may have seemed to the founding fathers of M.A.A.—to justify two separate organizations of mathematicians, but I would claim that it provides no such justification, especially today. For it is simply untrue that the faculty of four-year colleges have no interest in research, as I, along with many others, can attest from personal experience. Indeed, it is today largely fortuitous whether a young mathematician finds her (or his) first employment in a university or college; and it is awful to think that this act of fate seals the mathematician into place on one side or the other of the divide.

I wish to underline the importance of *not* classifying mathematicians according to whether they teach at university or college. I recall participating in a small conference at Carleton College in the summer of 1984, when the idea was being mooted of founding an institute for the problems of mathematics education. A view that emerged at that conference was that such an institute would be of interest to college mathematicians but not to university mathematicians, and that one should therefore seek to create a consortium of colleges to act as an umbrella organization. I totally reject that view, which I believe betrayed the existence of a superiority complex among the university mathematicians, and an inferiority complex among the college mathematicians propounding it. If asked to speak at the mathematics colloquium at a four-year college, I would regard it as insulting to my hosts not to talk on a piece of interesting research, provided I had such a tasty morsel available. Of course, I would make a real effort to make my talk intelligible to a general audience of mathematicians, with respect to both motivation and detailed content, but so, I should hope, would any speaker at any colloquium, and the task would not be substantially different at Carleton College or Cornell University.

In pursuing my researches into the historical origins of the unfortunate split between A.M.S. and M.A.A., I have looked for other examples of a separation of those who preach from those who practice. I have discovered a precedent. Some time between Euclid and Archimedes a religion arose, and gained many adherents, in which a group of devotees were designated as priests or preachers and drew up sets of rules for those who were to practice certain mystic rites of human communion. I remark that the preachers did not themselves practice these rites and that their precepts tended to be negative and inhibitory. Further, the preachers by no means can be said to have encouraged new ideas, innovation and experiment, and were fundamentally opposed to the introduction of new technology. I conclude that the model provided by this ancient religious institution should probably not have been adopted in seeking to design an appropriate modern structure for mathematical progress.

3. The Old Controversy, Morris Kline and Max Newman.

Of course, the debate as to whether teaching and research are complementary or antithetical has geen going on for some time. That they are indeed naturally antithetical and not simply in conflict due to the application of crude sociological forces such as the predilections of university administrations has been eloquently argued by Morris Kline in his well-known book *Why the Professor Can't Teach* [8]. I endeavored at the time, in my review of [8] which appeared in the *American Mathematical Monthly* [3], to rebut Kline's arguments and I do not propose to repeat those arguments here. I only wish to emphasize again why I do not believe the conflict to be natural in the sense that it inheres, unavoidably, in the nature of our responsibilities and activities as university mathematicians. I argue, as I did in [4], that "the only valid criterion of effective teaching we have, however unreliable, is a love of one's subject and (that) the characteristic way in which a mathematician expresses that love is by engaging in research."

Responding to my article in a letter to *Focus* [1], Ralph Boas disputed the second half of the above proposition[1], and asked me to replace the word "research" by "scholarship." He thus joins Morris Kline in entering a plea for the role of the scholar rather than the researcher, arguing that the scholar combines excellent judgment and maturity with a genuine love of teaching. I do not dispute that scholars, so characterized, would be invaluable members of mathematics faculties, in universities or colleges; I simply claim that we have no idea how to produce such people! They are to be compared to genetic sports; one cannot base one's educational strategy on the chance emergence of such mutations. We do know how to produce research mathematicians and are very successful at doing so. The pool of people who apprentice themselves to our trade consists predominantly of those who enjoy mathematics; and those who complete their apprenticeship and become research mathematicians are, broadly speaking, those who retain that enthusiasm.

Boas also argues that research and teaching can interfere with each other—one may neglect one's teaching when wrapped up in a piece of research, or one may neglect one's research when preparing a new set of lectures. Of course this is true; but it is only a special case of the proposition that there can be interference whenever one's job, or one's art, has more than one facet. The researcher who does not teach may well find a conflict between the claims of two pieces of research or between working

---

[1] Boas wrote "I fear that 'love is enough' is no more valid in mathematics than in marriage." The quotation marks round 'love is enough' might lead the reader to suppose I had asserted this.

on research and writing a book; likewise the teacher may have to weigh the claims on his attention of a large course containing familiar material and a smaller, more specialized course involving new material. I will, however, argue below that, if the term "research" is interpreted more liberally, then my proposition on the harmonious interplay of teaching and research becomes irrefutable. Of course, it would be unfair to adopt this generalized meaning of research in my dispute with Morris Kline, since he does not understand research in this broader sense.

It seems, too, that I must again (and again!) repudiate the position that there exists *de facto* no conflict between teaching and research. I agree with both Kline and Boas that there does, and that university administrations do tend to set too much value on published research—it is very difficult to educate university administrators. In the blurb on the dustjacket of Kline's book we find "The ostensible cause of the poor undergraduate education is the insistence by most university and college administrations on strength in research as the sole qualification for any professional status and tenure." The quotation is puzzling since the primary meaning of "ostensible" is "apparent, seeming" with the connotation that the appearance is misleading (thus our ostensible purpose, in supporting the Contras, is to restore democracy to Nicaragua); while the secondary meaning is "obvious, self-evident." Is Kline's ostensible purpose in this quotation to attack both administrators and research mathematicians simultaneously?

The opposite view to that of Kline was splendidly exemplified by the British topologist M. H. A. (Max) Newman, to whom I am personally indebted for so much, including my conviction of the complementarity of teaching and research. I had worked in a section of military intelligence closely allied to that run by Newman while we were engaged on breaking German codes during World War II and I had come to admire his administrative flair, not realizing at the time how rare this talent was. After the war, after two years at Oxford as Henry Whitehead's research student, I went to Manchester in 1948 to take up my first university appointment in the mathematics department of the university, headed by Max Newman (pure) and Sydney Goldstein (applied). Newman was in the process of building up one of the finest groups of research mathematicians in the country—among his appointments one finds such illustrious names as Graham Higman, Bernhard Neumann, J. W. S. Cassels, Paul Cohn, David Rees, to name but a few—and thereby establishing the reputation of his department and the quality of its output of original work. But Newman recognized, too, the importance of good and conscientious teaching and introduced certain procedures to raise the quality of the instruction—procedures for which he received the full support of his colleagues, even though they involved more work! Thus syllabi were published for all the courses and agreed by the entire department; and examination papers (with each question accompanied by a model answer) were also scrutinized and debated at a full meeting of the faculty. The two aspects of our professional duties, teaching and research, were harmoniously linked under Newman's quietly effective guidance, to the great advantage of both—and of ourselves.

4. The Report of the Panel of The Association of American Colleges.

I would not have thought it either necessary or appropriate to devote so much attention to this "pseudo-conflict" were it not that the recent appearance of an influential report left me seething with anger and disgust—indeed it was this report which decided me as to the topic of my address when I was invited by Ken Gross to speak at this conference. I have already responded to the report [2] in an article in *Focus* [4], but, in view of the somewhat ephemeral nature of that journal, I ask your indulgence for repeating my case here.

The A.A.C. Report of the *Project on Redefining the Meaning and Purpose of Baccalaureate Degrees* was issued in February, 1985. It represents the result of the deliberations of a 19-member committee set up in 1982. My first complaint is that there was not a single mathematician on the committee! How can a group of people deliberate for three years on a crucial issue of undergraduate education without having among them even one representative of what is arguably the most important discipline in the undergraduate curriculum[2]? How one sighs for the days when education was clearly distinguished from training, so that one mark of the educated person was that that person knew what she (or he) did not know.

Thus the conclusions of this august committee are discredited from the outset; it is no wonder that their diagnosis of the problem—and there is no denying there is a problem—is so wide of the mark. The following quotation epitomizes their position—and their recommendations:

> Central to the troubles and to the solution are the professors ... the development that overwhelmed the old curriculum and changed the entire nature of higher education was the transformation of the professors from teachers concerned with the characters and minds of their students to professionals, scholars with Ph.D. degrees with an allegiance to academic disciplines stronger than their commitment to teaching or to the life of the institutions where they are employed.

Whether this diagnosis is rubbish in another discipline, say English, I cannot say with certainty—though doubtless the historians, philosophers and biologists on the committee could. But I state categorically that, applied to mathematics, it could not be wider of the mark. It is obvious, and will again be obvious during the

---

[2] The policy of ignoring mathematicians is widespread in the academic community. I have before me an invitation to a conference at the Center for Tomorrow (!), SUNY, Buffalo, "to study the relation between liberal arts and technology courses." Of 14 invited speakers, not one is a mathematician!

panel discussions tomorrow, that active mathematicians have a keen concern for the curriculum and a commitment to good teaching. I have often marvelled—to quote from [4]—" ... with what care and concern, and at what a cost in time and energy and effort, each of us, in our own institutions, endeavors to improve the curriculum and render it more accessible and more appropriate to our students' needs, interests and talents without sacrifice of integrity—and this despite the convincing evidence of the invariance of student response to such change in the curriculum." I do not know the committee's evidence of a weakening of faculty commitment to the life of their institutions. If I am to judge by the desire for tenure of untenured faculty or by the statistic of the permanence of tenured faculty, I would surely reach the opposite conclusion!

The committee endeavored to support their diagnosis of the malaise by looking at graduate studies, and detected a significant neglect of any preparation of the future teacher: " ... during the long years of work towards the doctoral degree, the candidate is rarely, if ever, introduced to any of the ingredients that make up the art, the science, and the special responsibilities of teaching ... yet the major career option for most holders of the Ph.D. degree is full-time teaching in a college or university."

What a farrago of misrepresentation we find here! First, there *is* no science of teaching and the authors of the report discredit themselves by referring to this mythical entity. It is precisely for this reason that it is so difficult to design a detection and reward system for good teaching comparable in reliability with that employed for good research. Second, it is untrue that most of our Ph.D.'s become full-time teachers—they become full-time faculty members expected to devote themselves to teaching, research—and committee work. Third, the implication that the graduate student neglects, and is encouraged to neglect, his future teaching duties is an unfair and unwarranted canard. Certainly she (or he) is encouraged to give her major attention to her thesis—and this is as it should be. But, more than ever before, the graduate student is made alert to the fact that teaching *is* important in itself, and that future judgments of her worth and effectiveness will take into consideration teaching skills and dedication. What our illustrious committee has not understood is that such skill and dedication is very unlikely to be found in one who is not devoted to her discipline—in our case, mathematics. Let us picture the scene—the candidate for a faculty position is asked the apparently innocent question "Do you love mathematics?" Not suspecting the hidden trap, the candidate replies with an enthusiastic affirmative. "In that case," he is told, "we regret we cannot offer you the position. We're looking for somebody who would make a good teacher." Does this not strike our 19 experts as ridiculous, as it surely strikes us?

5. A Broadened View of Research.

Mathematical talent may be shown in various ways, of which teaching and research in the traditional sense of creating new mathematics are but two. I recall the time when applied mathematicians were often at a disadvantage on the faculty of a mathematics department because their "pure" colleagues would argue that finding new applications of existing mathematics did not increase our knowledge of mathematics itself. I do not believe many would be found today to support this extreme—and absurd—view that research on new applications is not mathematical research. I would myself plead for an even broader view of what constitutes research in order to bring into even closer relation the nature of the appropriate activities of university and college mathematicians and also to include Boas' scholar as more than a teacher. It is especially true today, as this conference and its predecessor [9] so unmistakably testify, that we must concern ourselves with curricular reform, for our majors, for our graduate students, and for those who study the mathematics requisite for their own discipline. The design and delivery of new curricula and new courses should surely be regarded as of a significance comparable to that of important new research, and the rewards should, I claim, be commensurable.

The accent should be on innovation. It is important to teach conscientiously and intelligibly but I do not believe that is enough to warrant promotion or tenure. To justify these one must innovate—and innovate successfully, whether it be within mathematics itself, in the applications of mathematics, in the development of new curricula or, for example, in the use of new technology for the delivery of instructional material. To borrow from the language of singularity theory, the "pure" teacher can at best produce continuous change; we should be on the lookout for those who can and do produce "catastrophes." If we introduce a bifurcation into Thom's phraseology, perhaps we should say that those who produce "anastrophes" should be rewarded with promotion and tenure, and those who produce "catastrophes" should be encouraged to seek fame and fortune outside the academic profession.

My final plea is of a different nature, but closely related to my main theme. Some of my colleagues argue that, while engaging in research may well invigorate one's presentation of graduate and even upper-division courses, it cannot have such an effect on the teaching of elementary courses—and may indeed there have a negative effect. I want to respond, a little obliquely, to this argument by claiming that the mathematician should be able to find scope for research, in the general sense, in even the most elementary mathematics. Let me end my remarks by giving two examples, taken from material appropriate even to a pre-college level.

In my first example, the mathematical problem first appeared when Jean Pedersen and I were engaged in trying to explain elementary ideas of probability theory to students aiming to become elementary teachers. We observed that, if two dice are thrown, the probability that the sum is divisible by 2 is $1/2$; that it is divisible by 3 is $1/3$; that it is divisible by 4 is $1/4$. But the probability that it is divisible by 5 is not $1/5$. From those simple empirical beginnings, we came to formulate the following question. Let x_i, $i = 1, 2, \ldots, k$, be independent random variables drawn from the rectangular distribution of integers $1, 2, \ldots, n$. We seek the relation between m, n, and k such that $\mathrm{prob}\left(\sum_{i=1}^{k} x_i \equiv 0 \bmod m\right) = 1/m$. It is easy to see that sufficient for this is $m|n$, and necessary is $m|n^k$. We conjecture that the property holds if and only if $m|n$ or there exist integers p, q such that $m = p^k$, $n = pq$ and $q \equiv -1 \bmod p^{k-1}$. We have succeeded in proving this for $k \leq 3$ [5].

A second attractive, and even more natural, combinatorial problem arises in connection with a procedure which Jean Pedersen and I have described [6], [7] for constructing arbitrarily good approximations to regular star polygons. This procedure was invented to try to revitalize the teaching of geometry at the junior high school and high school levels. Our systematic procedure led, by a very natural, but not entirely obvious, generalization to the following algorithm for determining the *quasi-order* of t mod b; that is, we determine the smallest integer k such that $t^k \equiv \pm 1 \bmod b$; and our algorithm also determines which sign to take. We understand that computer scientists have some interest in this problem. I now describe the algorithm, *first assuming that t is odd*.

Thus, let t, b be mutually prime with t odd and let $S = \{a, \text{ such that } 0 < a < b/2,$ $t \nmid a\}$. We then claim that there exists exactly one q in $0 < q \leq (t-1)/2$ such that $qb + (-1)^\epsilon a$ is divisible by t with $\epsilon = 0$ or 1. Let us then write

$$qb + (-1)^\epsilon a = t^k a', \qquad \text{with } k \text{ maximal.} \tag{5.1}$$

We next claim that the function $a \mapsto a'$, given by (5.1) is a permutation of S. Thus if we commence with a given a_1 in S we obtain a sequence of equations

$$q_i b + (-1)^{\epsilon_i} a_i = t^{k_i} a_{i+1}, \qquad i = 1, 2, \ldots, r; \; a_{r+1} = a_1 \tag{5.2}$$

We record (5.2) in the symbol

$$b \begin{vmatrix} a_1 & a_2 & \ldots & a_r \\ k_1 & k_2 & \ldots & k_r \\ \epsilon_1 & \epsilon_2 & \ldots & \epsilon_r \end{vmatrix} \tag{5.3}$$

with, by definition,

$$k = \sum k_i, \qquad \epsilon = \sum \epsilon_i, \tag{5.4}$$

It is easy to see from (5.2) that $\gcd(b, a_i)$ is independent of i; and we call the symbol (5.3) *reduced* if $\gcd(b, a_i) = 1$. We describe (5.3) as *contracted* if the a_i are all distinct, that is, if $a_1, a_2, \ldots, a_r$ are permuted cyclically by the function $a \mapsto a'$ described in (5.1). Then our result is the following:

Theorem. *If the symbol (5.3) is reduced and contracted, then the quasi-order of t mod b is k; and, indeed, $t^k \equiv (-1)^\epsilon \bmod b$. (Here k and ϵ are defined by (5.4)).*

If t is even, the only modification is in the range of values of q for which we consider the expression $qb + (-1)^\epsilon a$. We then consider this expression for $\epsilon = 0$,

$0 < q \leq t/2 - 1$; and for $\epsilon = 1$, $0 < q \leq t/2$. The statement of the theorem remains the same.

It is pleasing to me (and to Jean Pedersen) that I can offer two examples of "computational mathematics" arising from such elementary sources.

6. Conclusion.

Let me end by repeating my tribute to the example which Gail Young has set for us all by his multi-faceted contribution to academic life as mathematician, scholar, administrator and teacher. His friendship has been, and continues to be, of immense value to me. It is a testimony to his magnanimity (if not to his judgment) that this friendship has even withstood the strain imposed by my having encouraged him to take a position at a certain institution whose anonymity I shall, as a gentleman, respect.

References.

[1] R. P. Boas, Letter to *Focus*, September 1985.
[2] *Chronicle of Higher Education*, February 13, 1985.
[3] P. Hilton, Review of *Why the Professor Can't Teach*, *Amer. Math. Monthly*, **86**(1979), 407–412.
[4] P. Hilton, "A job on our hands," *Focus*, May–June 1985.
[5] P. Hilton and J. Pedersen, "On the distribution of the sum of a pair of integers," *Spectrum*, **20**(1982), 29–36.
[6] P. Hilton and J. Pedersen, "Folding regular star polygons and number theory," *Math. Intelligencer*, **7**(1985), 15–26.
[7] P. Hilton and J. Pedersen, "On certain algorithms in the practice of geometry and the theory of numbers," *Publicacions Sec. Mat.*, *U.A.B.*, 29, 1(1985), 31–64.
[8] M. Kline, *Why the Professor Can't Teach*, St. Martin's Press, 1978.
[9] P. Hilton and G. S. Young, Jr. (eds.), *New Directions in Applied Mathematics*, Springer, 1982.
[10] P. Hilton, "Education in mathematics and science today: the spread of false dichotomies," *Proc. ICME III*, (1976), 75–97.

Peter Hilton was born in London, England, on April 7, 1923. He was educated at Oxford University and spent the years 1942–45 of the Second World War as a cryptanalyst at Bletchley Park, where he became very friendly with J. H. C. (Henry) Whitehead. On demobilization he returned to Oxford to obtain a D.Phil. in algebraic topology under Whitehead's supervision. After an academic career in England, culminating in his holding the Mason Chair of Pure Mathematics at the University of Birmingham, he came to the U.S. in 1962 to become Professor of Mathematics at Cornell University. He is now Distinguished Professor of Mathematics at SUNY-Binghamton. Hilton's research interests are in algebraic topology and homological algebra, but he has taken an increasing interest in mathematics education, at all levels, since coming to the U.S. He has been Chairman of the U.S. Commission on Mathematical Instruction and Secretary of the International Commission. He was Chairman of the NRC Committee on Applied Mathematics Training, 1977–79.

STOCHASTIC POPULATION THEORY:
MATHEMATICAL EVOLUTION OF A GENETICAL MODEL

KENNETH J. HOCHBERG

Department of Mathematics and Statistics
Case Western Reserve University
Cleveland, Ohio 44106

Dedicated to Gail Young
on his seventieth birthday

1. Introduction.

The use of mathematical modeling in characterizing and analyzing stochastic fluctuation phenomena among individuals in natural populations has a long and honorable history dating back to the first quarter of the twentieth century. Indeed, more than sixty years have passed since R. A. Fisher (1922) introduced diffusion theory in his analysis of the dominance ratio in population genetics theory. However, in the last decade, coincident with monumental progress in computational techniques and the introduction of novel ideas from the mathematical theory of probability and stochastic processes, the pace of the evolution of stochastic population theory—the use of mathematical techniques to explain and predict the random behavior of individuals in natural populations—has quickened markedly.

In this paper, we will follow the development over the past decade of the mathematical study of a particular model describing the allelic frequency distribution of a selectively neutral gene at a single locus on a chromosome of individuals in some large fixed-size randomly mating natural population. The intent of the paper is not so much to derive all the theorems or even to characterize the fluctuations in the behavior of such genetic populations; other papers in both the biological and mathematical literature do this quite adequately. Rather, we will attempt to follow the evolution of the mathematics applied to the problem: the development of new mathematical concepts and techniques motivated by the genetical problem and their incorporation into the analysis of the model. These mathematical ideas, motivated as they were by biological considerations, have now been applied to several other areas and have become the objects of study of theoretical mathematicians in their own right.

2. The Basic Model: Discrete Time, Discrete State-Space, One Dimension.

We consider first a natural population of N individuals and assume that the population size is large and remains fixed, perhaps because of limitations due to finite natural resources or limited space. For simplicity, we will assume that the individuals are haploid; thus, each member of the population has only one gene at each locus on each chromosome. (Diploid individuals have two genes at each locus; one can place them into the context of our model by considering the two genes as independent objects—call them gametes—and studying the frequency distribution of the gametic population.) The various genetic types are called alleles, and the objective of the model is to describe the allelic frequency distribution of the population at time t.

Underlying this entire discussion is what Ewens (1979) has referred to as "the leading problem of induction in population genetics theory," that of fitting observations of the evolution of a genetical system to some hypothesis of genetic evolution. To do this, one must first deduce what might be expected under the circumstances of one's hypothesis; hence, the need for a reasonable mathematical model of the fluctuations in the allelic frequency distribution. Ultimately, the process of modeling to fit a hypothesis and, then, using acceptable models to predict future behavior is one that involves both deductive and inductive considerations. It requires, fundamentally, the simultaneous application of biological, statistical, and mathematical concepts.

Kingman (1980) has pointed out that it is now widely accepted that the evolution of natural populations is to be understood in terms of Mendelian genetics and Darwinian natural selection. In other words, one inherits genetic properties from one's parents; however, some genetic types have a selective advantage over others, and "the fittest" have a greater probability of survival. Nonetheless, tremendous variability is clearly evident in most natural populations, including our own human race. Does this not conflict with Darwinian theory? At what rate should biologists expect evolution to occur? One can hypothesize that selective pressure can maintain a balanced polymorphism when heterozygotes are sufficiently favorable. In addition, even when selective effects tend to reduce diversity in the genetic population, the opposite tendency of mutation to create new genetic types might bring all into balance. Perhaps selection favors different alleles in different regions, and migration between regions creates diversity everywhere. Such are the types of hypotheses that geneticists wish to fit to mathematical models.

One technique that had been used until recently for classifying allelic types is electrophoresis, by which the electrical charge of each protein is measured. It is possible that two proteins differing by one or more amino acids may have the same charge, so not all (in fact, according to Ewens (1979), perhaps only about one-third of) genetic variation can be so resolved. Ohta and Kimura (1974) introduced their so-called "ladder" or "charge-state" or "stepwise-mutation" model to describe the electrical charge states of alleles in a population of N haploid individuals as a discrete-time Markov process $\{N(t) : t = 0, 1, \ldots\}$ with transition probabilities describing Wright-Fisher-type multinomial sampling of N individuals at random according to the empirical distribution of the parent or host population, resulting in random genetic drift, i.e.,

$$P\left\{ \begin{array}{c} D_i \text{ genes of allelic type } A_i \text{ in generation } t+1, \\ \text{given } N_i \text{ genes of type } A_i \text{ in generation } t, \\ (i = 1, \ldots, k) \end{array} \right\}$$

$$= \frac{N!}{D_1! \, D_2! \cdots D_k!} \, (N_1/N)^{D_1} \, (N_2/N)^{D_2} \cdots (N_k/N)^{D_k}, \quad \sum_{i=1}^{k} D_i = N. \tag{2.1}$$

In addition, since mutations ought to be viewed as small deviations from the natural Mendelian structure, they assume that a mutation consists of a change in the charge of a single amino acid and, correspondingly, a change in the charge-state of one unit. Thus, in the Ohta-Kimura model, one superimposes upon the Markov chain description given above the stipulation that, at the moment of birth, an individual born of allelic type A_i can mutate to type A_{i-1} or type A_{i+1}, each with a small given mutation probability $u/2$.

It should be noted that we are not now incorporating a Darwinian selection feature into the model. This is consistent with the so-called neutral theory of Kimura (1968), that whereas the gene substitutions responsible for obviously adaptive and progressive phenomena are clearly selective, many other gene substitutions have occurred purely by chance through stochastic processes acting in finite populations, and for these the replacing gene has no selective advantage over its predecessor, or, indeed, over any other extant genetic variation.

3. The Basic Model: Continuous Time, Discrete State-Space, Several Dimensions.

We now consider a generalization of the Ohta-Kimura ladder model to continuous time and several dimensions. Now, the state of an individual no longer represents its electrical charge but is, rather, a vector $k = (k_1, k_2, \ldots, k_d)$ describing d observable integer-valued traits of the individual, such as height, weight, or length of wing-span. Let $\mathbf{Z}$ be the set of integers, and let $N_k(t)$ denote the number of individuals of type $k \in \mathbf{Z}^d$ in the population at time $t \geq 0$. Let $M_1(\mathbf{Z}^d)$ be the set of probability measures on $\mathbf{Z}^d$. For $t \geq 0$, define $p(t, \cdot) \in M_1(\mathbf{Z}^d)$ by

$$p(t; k) \equiv N_k(t)/N$$
$$= \text{empirical frequency of type } k \text{ at time } t, \tag{3.1}$$

for $k \in \mathbf{Z}^d$. Then, $\{p(t, \cdot) : t \geq 0\}$ is an $M_1(\mathbf{Z}^d)$-valued continuous-time Markov process with generator given by

$$L_N^1 f(p) = \sum_{i \neq j} \left[\gamma p(i)p(j) + Dp(i) \sum_k p(k)\theta_{kj} \right] \left[f(p^{ij}) - f(p) \right] \qquad (3.2)$$

where $\gamma > 0$ represents the sampling rate, $D > 0$ is the mutation rate, $p \in M_1(\mathbf{Z}^d)$, $f \in C_b\big(M_1(\mathbf{Z}^d)\big)$, the space of bounded continuous functions on $M_1(\mathbf{Z}^d)$,

$$p^{ij}(k) = \begin{cases} p(k) - 1/N & \text{if } k = i \\ p(k) + 1/N & \text{if } k = j \\ p(k) & \text{if } k \neq i, j \end{cases} \qquad (3.3)$$

and

$$\theta_{ij} = \begin{cases} 1 & \text{if } |i - j| \equiv \sum_{k=1}^d |i_k - j_k| = 1 \\ -2d & \text{if } i = j \\ 0 & \text{otherwise.} \end{cases} \qquad (3.4)$$

The evolutionary mechanism here is as follows: at a constant rate, an individual of type i is replaced at the time of his death by one of parent type k and is subject to a possible mutation from k to j. One can alter the mechanism to permit mutation between types at any time, not necessarily coincident with the death of one individual and the simultaneous birth of another, by replacing $L_N^1 f(p)$ with

$$L_N^2 f(p) = \sum_{i \neq j} \left[\gamma p(i)p(j) + Dp(i)p(j) \right] \left[f(p^{ij}) - f(p) \right]. \qquad (3.5)$$

According to the latter formulation, if an individual dies, it is always assumed to be the parent of its replacement.

Early analysis of the model included the observation of Moran (1975, 1976) that the Markov process has no stationary distribution; he therefore referred to the allelic frequency distribution as a "wandering" one. Kingman (1976) followed with the observation that the joint distribution of relative differences of the locations of states measured from the location of any specific randomly chosen allele does converge as t becomes infinite, though his existence proof is non-constructive and does not provide a description of the limiting joint distribution. Kingman referred to the sequence of generations in the discrete-time model as a "coherent random walk," since the allelic distribution tends to cluster. These results imply that for a fixed finite population size N, there is a limiting random number of types as t becomes infinite. Kesten (1980a,b) proved that if the mutation rate decreases at a rate that is inversely proportional to the population size N, then the limiting number of types become infinite with N, but at an extremely slow rate. This result had not been at all evident from the earlier Monte Carlo simulations of Kimura and Ohta (1975, 1978).

4. The Measure-Valued Diffusion Approximation.

Fleming and Viot (1978, 1979) introduced an alternative limiting form of the Ohta-Kimura ladder model, which can be viewed as a diffusion approximation in the case of small incremental mutational effect. Specifically, in contrast to the analyses described in the last section, in which the limiting behavior of the allelic distribution was obtained under the assumptions that the mutation rate is inversely proportional to the population size N and that the incremental effect of a single mutation on the numerical characteristics of an individual remains constant, Fleming and Viot assume that the mutation rate D remains constant, but that the incremental effect of a single mutation decreases at a rate that is inversely proportional to the square root of the population size. That is, the Ohta-Kimura Markov process $\{p(t, \cdot) : t \geq 0\}$ is rescaled in both space and time to yield an $M_1(S)$-valued stochastic process

$$Y_N(t, A) \equiv \sum_{N^{-1/2} j \in A} p(N^2 t, j), \tag{4.1}$$

where $M_1(S)$ denotes the space of probability measures on $S = \mathbb{R}^d \cup \{\infty\}$, the one-point compactification of $\mathbb{R}^d$, furnished with the topology of weak convergence of probability measures.

If one begins with the Ohta-Kimura process given by the generator in (3.2), then the resulting process $Y_N(\cdot)$ is a Markov process with state space $M_1^N(\mathbb{R}^d)$, the space of probability measures concentrated on the lattice $N^{-1/2} \mathbb{Z}^d$ and consisting of the superposition of N atoms of mass $1/N$. The generator L_N^1 is given by

$$
\begin{aligned}
L_N^1 \psi(p) &= \sum_{i,j} N^2 \big[f\big(\langle \phi, p \rangle - \phi(x_i)/N + \phi(x_j)/N \big) - f(\langle \phi, p \rangle) \big] \\
&\qquad \cdot \Big[\gamma p(i)p(j) + Dp(i) \sum_k p(k)\theta_{kj} \Big] \\
&= N^2 Df'(\langle \phi, p \rangle) \Big\{ \sum_i p(i) \sum_k p(k) \Big[\sum_{|k-j|=1} \phi(x_j) - 2d\phi(x_k) \Big] /N \Big\} \\
&\quad + \tfrac{1}{2} \gamma f''(\langle \phi, p \rangle) \Big\{ \sum_{i,j} p(i)p(j) \big[\phi^2(x_i) - 2\phi(x_i)\phi(x_j) + \phi^2(x_j) \big] \Big\} \\
&\quad + R_1(\phi)/N^{1/2},
\end{aligned}
\tag{4.2}
$$

where the domain of L_N^1 is the linear space of functions of the form

$$
\begin{aligned}
\psi(p) &= f(\langle \phi, p \rangle) \qquad \text{for } f \in C_b^3(\mathbb{R}^1),\ \phi \in C_b^3(\mathbb{R}^d), \\
\langle \phi, p \rangle &= \sum_i \phi(x_i)p(i), \qquad x_i = i/N^{1/2},
\end{aligned}
\tag{4.3}
$$

and $R_1(\phi)$ is a bounded remainder term if $|f'''| \leq M < \infty$ for some constant M. If one uses the Ohta-Kimura generator given by (3.5), then the rescaled process Y_N

has generator L_N^2 given by

$$
\begin{aligned}
L_N^2 \psi(p) = &\sum_{i,j} N^2 \big[f(\langle \phi, p \rangle - \phi(x_i)/N + \phi(x_j)/N) - f(\langle \phi, p \rangle) \big] \\
&\quad \cdot \big[\gamma p(i) p(j) + Dp(i) \theta_{ij} \big] \\
= &\, N^2 Df'(\langle \phi, p \rangle) \Big\{ \sum_i p(i) \Big[\sum_{|i-j|=1} \phi(x_j) - 2d\phi(x_i) \Big] / N \Big\} \\
&\quad + \tfrac{1}{2} \gamma f''(\langle \phi, p \rangle) \Big\{ \sum_{i,j} p(i) p(j) \big[\phi^2(x_i) + \phi^2(x_j) - 2\phi(x_i)\phi(x_j) \big] \Big\} \\
&\quad + R_2(\phi)/N^{1/2},
\end{aligned}
\tag{4.4}
$$

where $R_2(\phi)$ is a bounded remainder term that differs slightly from $R_1(\phi)$. In either case, one obtains a generator of the form

$$
\begin{aligned}
L_N \psi(p) = &\, Df'(\langle \phi, p \rangle)(\langle \Delta \phi, p \rangle) \\
&\quad + \gamma f''(\langle \phi, p \rangle) \big[\langle \phi^2, p \rangle - \langle \phi, p \rangle^2 \big] \\
&\quad + R(\phi)/N^{1/2},
\end{aligned}
\tag{4.5}
$$

where Δ denotes the d-dimensional Laplace operator.

Fleming and Viot identified the weak limit of this process as N increases as a Markov diffusion process with generator given by

$$
\begin{aligned}
L\psi(\mu) = &\, D \sum_{i=1}^{n} f_{y_i}\big(\langle \phi_1, \mu \rangle, \cdots, \langle \phi_n, \mu \rangle \big) \langle \Delta \phi_i, \mu \rangle \\
&\quad + \gamma \sum_{j=1}^{n} \sum_{i=1}^{n} f_{y_i y_j}\big(\langle \phi_1, \mu \rangle, \ldots, \langle \phi_n, \mu \rangle \big) \\
&\quad\quad \cdot \big[\langle \phi_i \phi_j, \mu \rangle - \langle \phi_i, \mu \rangle \langle \phi_j, \mu \rangle \big],
\end{aligned}
\tag{4.6}
$$

where the domain of L is the set of functions of the form

$$
\psi(\mu) = f\big(\langle \phi_1, \mu \rangle, \ldots, \langle \phi_n, \mu \rangle \big)
\tag{4.7}
$$

for $\phi_i \in C_K^2(\mathbb{R}^d)$, the space of twice continuously differentiable functions on $\mathbb{R}^d$ with compact support,

$$
\langle \phi, \mu \rangle \equiv \int \phi(x)\,\mu(dx), \qquad f(y_1, \ldots, y_n) \in C^2(\mathbb{R}^n),
\tag{4.8}
$$

and f_{y_i} and $f_{y_i y_j}$ denote the first and second partial derivatives of f.

Operators associated with diffusion processes with values in $M_1(S)$ have the form

$$
L\psi(\mu) \equiv \int_S A\left(\frac{\delta \psi(\mu)}{\delta \mu(x)} \right) \mu(dx) + \int_S \int_S \frac{\delta^2 \psi(\mu)}{\delta \mu(x)\, \delta \mu(y)} Q(\mu : dx \times dy),
\tag{4.9}
$$

where

$$\frac{\delta\psi(\mu)}{\delta\mu(x)} \equiv \lim_{\epsilon\downarrow 0} \frac{\psi(\mu + \epsilon\delta_x) - \psi(\mu)}{\epsilon}, \tag{4.10}$$

$Q : M_1(S) \to M_1(S \times S)$ (quadratic fluctuation functional),

A is the infinitesimal generator of a strongly continuous Markov semi-group on $C_0(\mathbb{R}^d)$, the space of continuous functions on $\mathbb{R}^d$ that vanish at infinity, and δ_x represents a unit mass at the point x.

The Fleming-Viot operator L given by (4.6) is of the form (4.9) with

$$A = D \cdot \Delta \tag{4.11}$$

and

$$Q(\mu : dx \times dy) = \gamma[\mu(dx)\,\delta_x(dy) - \mu(dx)\,\mu(dy)]. \tag{4.12}$$

If we denote by $X(\cdot, \cdot)$ the $M_1(S)$-valued Markov diffusion process given by the Fleming-Viot operator L on its domain, then we have the following result:

Theorem 4.13 (Fleming and Viot (1979)). *Let $Y_N(\cdot, \cdot)$ be given as in (4.1). Then, $Y_N(\cdot, \cdot) \overset{w}{\to} X(\cdot, \cdot)$ as $N \to \infty$, where $\overset{w}{\to}$ denotes convergence in the sense of weak convergence of probability measures on $\Omega^D \equiv D([0, \infty), M_1(S))$, the space of functions mapping $[0, \infty)$ into $M_1(S)$ that are right continuous with limits from the left.*

5. The Function-Valued Dual Process.

Spitzer (1970) introduced the notion of dual processes for studying the distribution of systems of infinitely many interacting particles. The technique was further developed by Holley and Liggett (1975) and by Holley and Stroock (1979) and was applied to the study of a class of finite-dimensional diffusion processes in Holley, Stroock and Williams (1977). Dawson and Hochberg (1982, 1985) applied the method of dual processes to measure-valued processes, and, specifically, they used the function-valued process dual to the Fleming-Viot model to obtain support, limiting, and scaling properties of the measure-valued process and, by extension, deduced qualitative properties of the original genetical model of Ohta and Kimura.

Let S be the one-point compactification of a locally compact complete separable metric space, and let S^N denote the N-fold Cartesian product of S. Let $\mathcal{C} = \cup_{N=0}^{\infty} C(S^N)$ furnished with the topology given by the inductive limit of the sup norm topologies, and let $C(S^0) = \mathbb{R}^1$. For $f \in \mathcal{C}$, define the function

$$N(f) = N \quad \text{if } f \in C(S^N). \tag{5.1}$$

Similarly, let $\mathcal{D} = \cup_{N=0}^{\infty} D(S^N)$ where, for each N, $D(S^N)$ is a dense subset of $C(S^N)$.

A function of the form

$$
\begin{aligned}
F_f(\mu) &= \int_S \cdots \int_S f(x_1, x_2, \ldots, x_{N(f)}) \, \mu^{(N(f))}(dx) \\
&\equiv \int_S \cdots \int_S f(x_1, x_2, \ldots, x_{N(f)}) \, \mu(dx_1) \cdots \mu(dx_{N(f)})
\end{aligned}
\tag{5.2}
$$

for $f \in C$ is called a **monomial** on $M(S)$, the space of bounded Radon measures on S. Let $\Pi_1(E)$ denote the set of monomials restricted to a compact subset E of $M(S)$, and let the **algebra of polynomials** $\Pi(E)$ be the smallest algebra of functions on E which contains $\Pi_1(E)$. Similarly, denote by $\Pi_1^D(E)$ and $\Pi^D(E)$ the monomials and polynomials on E with coefficients $f \in D$. Let $\mathcal{Q}$ be the linear space of functions on D which contains functions of the form

$$
F(\mu, f) = F_\mu(f) \equiv F_f(\mu) \qquad \text{for } \mu \in E, \ f \in D.
\tag{5.3}
$$

Now consider a pregenerator L_1 with domain $D(L) \subset \Pi^D(E)$ such that for $F_f \in D(L)$, L_1 can be related to another pregenerator L_2 via the relationship

$$
L_1 F_f(\mu) = L_2 F_\mu(f) + V\big(N(f)\big) F_\mu(f)
\tag{5.4}
$$

where V is a function defined on the non-negative integers.

A probability measure P_μ is said to be a **solution to the martingale problem** for a pregenerator L if for every pair (f, g) with $f \in D(L)$ and $g = Lf$,

$$
f\big(X(t)\big) - \int_0^t g\big(X(s)\big) \, ds
\tag{5.5}
$$

is a P_μ-martingale. P_μ is said to have initial condition μ if

$$
P_\mu\{X(0) = \mu\} = 1.
\tag{5.6}
$$

If the martingale problem associated with $\{(F_f, L_1 F_f) : f \in D(L_1)\}$ has an $M(S)$-valued solution $\{X(t) : t \geq 0\}$ and the martingale problem associated with $\{(F_\mu, L_2 F_\mu) : \mu \in E\}$ has a $\Pi^D(E)$-valued solution $\{Y(t) : t \geq 0\}$, then $X(t)$ and $Y(t)$ are said to form a pair of **dual processes**. It should be noted that it is possible for a measure-valued process to have several different function-valued duals.

A function-valued dual process to the Fleming-Viot model was described in Dawson and Hochberg (1982). Let $E = M_1(S)$, the set of probability measures on the one-point compactification of $\mathbb{R}^d$, and form the algebra of polynomials $\Pi\big(M_1(S)\big)$ as before. For $f \in C^\infty(S^N)$, the dual process has generator given by

$$
LF_f(\mu) = \int \cdots \int K f(x_1, x_2, \ldots, x_N) \, \mu(dx_1) \cdots \mu(dx_N)
\tag{5.7}
$$

where

$$Kf(x_1,\ldots,x_N) = D\sum_{j=1}^{N} \Delta_j f(x_1,\ldots,x_N)$$

$$+\gamma \sum_{j=1}^{N} \sum_{\substack{k=1 \\ k\neq j}}^{N} \big[\Phi_{jk}\big(f(x_1,\ldots,x_N)\big) - f(x_1,\ldots,x_N)\big]. \tag{5.8}$$

Here, Δ_j denotes the d-dimensional Laplacian operating on the variable x_j, and

$$\Phi_{jk} : C(S^N) \to C(S^{N-1}) \tag{5.9}$$

such that

$$\Phi_{jk}\big(f(x_1,\ldots,x_N)\big) = (\Phi_{jk}f)(y_1,\ldots,y_{N-1}) \tag{5.10}$$

where for $j \neq k$,

$$
\begin{aligned}
x_i &= y_i & \text{for} \quad & i = 1,\ldots,k-1 \\
x_k &= y_j & \text{if} \quad & j < k \\
x_k &= y_{j-1} & \text{if} \quad & j > k \\
x_i &= y_{i-1} & \text{for} \quad & i = k+1,\ldots,N.
\end{aligned}
$$

Equivalently,

$$LF_f(\mu) = F_{\Delta_{N(f)}f}(\mu) + \gamma \sum_{j=1}^{N(f)} \sum_{\substack{k=1 \\ j\neq k}}^{N(f)} \big[F_{\Phi_{jk}f}(\mu) - F_f(\mu)\big], \tag{5.11}$$

so we have the duality relationship

$$L_1 F_f(\mu) = L_2 F_\mu(f), \tag{5.12}$$

where $L_1 = L$ and L_2 is a linear operator defined on $\mathcal{Q}(\mathcal{C})$.

L_2 is itself the infinitesimal generator of a Markov process $\{Y(t) : t \geq 0\}$ on $\mathcal{C}$ with dynamics described by the following:

a) at rate $\gamma n(n-1)$, $Y(t)$ jumps from $C(S^N)$ to $C(S^{N-1})$ for $N \geq 2$, and, at the time of a jump, f is replaced by $\Phi_{jk}f$;

b) between the jumps described in a), $Y(t)$ is deterministic on $C(S^N)$ and evolves according to the Brownian motion semigroup H_t^N on $(\mathbb{R}^d)^N$; i.e.,

$$H_t^N f(x) = (2\pi t)^{-Nd/2} \int \exp\big(-|x-y|^2/2t\big) f(y)\,dy; \tag{5.13}$$

and

c) once $Y(t)$ is a function of only one variable, i.e., $Y(t) \in C(S)$, no further jumps occur.

The duality relationship

$$E_\mu\{F_f(X(t))\} = E_f\{F_\mu(Y(t))\},\tag{5.14}$$

where

$$
\begin{aligned}
E_\mu\{F_f(X(t))\} &= T_t F_f(\mu)\\
&= E_\mu\left\{\int \cdots \int f(x_1,\ldots,x_N)\,X(t,dx_1)\cdots X(t,dx_N)\right\}
\end{aligned}\tag{5.15}
$$

and

$$
\begin{aligned}
E_f\{F_\mu(Y(t))\} &= F_\mu(E_f\{Y_t\})\\
&= \int \cdots \int E_f\{Y_t(x_1,\ldots,x_{N(t)})\}\,\mu(dx_1)\cdots\mu(dx_{N(t)}),
\end{aligned}\tag{5.16}
$$

turns out to be a key element in analyzing the qualitative behavior and stochastic geometry of the limiting Fleming-Viot process $X(t)$.

6. Ergodic Behavior, Scaling Limits, and Support Properties.

As we stated in the introduction, the primary objective of this article is not to analyze the stochastic processes being discussed, but rather to follow the mathematical development of the analysis over the past decade. Therefore, we present in this section only a few of the results that have been obtained using the measure-valued approximations and function-valued duals that we have introduced in previous sections. More complete details and applications can be found in the series of papers by Dawson and Hochberg (1982, 1983, 1985, 1986).

Let the empirical centered moments of $\{X(t) : t \geq 0\}$ be defined by

$$R_{k_1,\ldots,k_d}(t) = \int_{\mathbb{R}^d} \prod_{i=1}^d (x_i - x_i(t))^{k_i}\,X(t,dx),\tag{6.1}$$

where the empirical mean process $x(t) \equiv (x_1(t),\ldots,x_d(t))$ is given by

$$x_i(t) = \int_{\mathbb{R}^d} x_i\,X(t,dx), \qquad i = 1,2,\ldots,d.\tag{6.2}$$

Then, the following result follows from observing the evolutionary nature of the dual process $\{Y(t) : t \geq 0\}$ as a jump process:

Theorem 6.3. If $N_0 = \sum_{\ell=1}^d k_\ell$ and $\int |x|^{N_0}\,\mu(dx) < \infty$, then

(i) $E_\mu\{R_{k_1,\ldots,k_d}(t)\} < \infty$ for $t \geq 0$,

    and

(ii) $\lim_{t\to\infty} E_\mu\{R_{k_1,\ldots,k_d}\} \equiv r_{k_1,\ldots,k_d}$ exists and is finite.

For the random cluster

$$X^*(t,dx) \equiv X(t,\{a - x(t) : a \in dx\})\tag{6.4}$$

centered at the empirical mean $x(t)$, we obtain the following description of its ergodic behavior from the dual process:

Theorem 6.5.

(i) X^* is a stationary process.

(ii)

$$I(f) = \lim_{t \to \infty} T^{-1} \int_0^T \int_{\mathbb{R}^d} f(x) \, X^*(t, dx) \, dt = \lim_{t \to \infty} E_\mu \left\{ \int_{\mathbb{R}^d} f(x) \, X^*(t, dx) \right\}$$

exists a.s. as a linear functional on $C(S)$ and is independent of the initial measure μ.

(iii)

$$\lim_{T \to \infty} T^{-1} \int_0^T X^*(t, A) \, dt = \nu(A)$$

$$= \text{expected value of the equilibrium random measure}$$
$$\text{on the set } A.$$

(iv) The set $\{r_{k_1, \ldots, k_d}\}$ forms the joint moment system of the expected steady-state distribution of the random cluster centered at the empirical mean.

It is worth noting that $\nu(A)$ is not necessarily a Gaussian measure.

For the rescaled process $\{X_\epsilon(t) : t \geq 0\}$ given by

$$X_\epsilon(t, A) = X(t/\epsilon^2; A_\epsilon), \qquad A_\epsilon \equiv \{x : \epsilon x \in A\}, \tag{6.6}$$

we have the following scaling limit result:

Theorem 6.7. *Assume $X(0)$ has compact support. Then, as $\epsilon \to 0$, the finite-dimensional distributions of the rescaled process $X_\epsilon(t)$ converge to those of the probability-measure-valued process $\delta_{w(t)}$ consisting of a single unit atom undergoing Brownian motion in $\mathbf{R}^d$.*

The Hausdorff-Besicovitch dimension of support of a Borel set E is defined by

$$\dim(E) = \sup \left\{ \beta > 0 : \lim_{\sigma \downarrow 0} \inf_{\mathcal{E}} \sum_i [d(E_i)]^\beta = \infty \right\} \tag{6.8}$$

where $d(E_i)$ is the diameter of the set E_i and

$$\mathcal{E} = \left\{ \{E_i\} : E_i \subset \bigcup_i E_i, \ d(E_i) < \sigma \text{ for each } i \right\}; \tag{6.9}$$

i.e., $\mathcal{E}$ is the set of all coverings of the set E by sets of diameter less than σ. A well-known result is that subsets of $\mathbf{R}^d$ of positive Lebesgue measure have Hausdorff dimension equal to d.

By tracing backwards in time the genealogy of an infinite system of particles determined via duality and the moment measures of the process $X(t)$, we can decompose the total mass at any time t into a set of genealogical chains in such a way that the surviving mass forms a generalized version of the classical Cantor set. Together with a technique due to Frostman, this leads us to the following results on the topological support of the random measure $X(t)$ at fixed times $t > 0$:

Theorem 6.10. *Assume $X(0)$ has compact support, with probability one. Then, the following hold:*

(i) *In any spatial dimension, $X(t, \cdot)$ has compact support, with probability one, for each fixed $t > 0$.*

(ii) *For $d \geq 2$ and $t > 0$, the Hausdorff-Besicovitch dimension of the topological support of $X(t)$ is equal to two.*

(iii) *$X(t, \cdot)$ is singular (i.e., not absolutely continuous with respect to Lebesgue measure) for $t > 0$ in $\mathbf{R}^d$, $d \geq 3$, with probability one.*

(iv) *$X(t, \cdot)$ is compactly coherent; i.e., there exists a random sphere $S(t, \omega)$ with radius $R(t, \omega)$ such that $R(t, \cdot)$ is a stationary stochastic process and such that $S(t, \omega)$ contains all the mass of the process $X(t, \cdot)$ with probability one, i.e.,*

$$P\{X(t, \omega, S(t, \omega)) = 1\} = 1. \tag{6.11}$$

Thus, the microscopic scale distribution of three or more genetic characteristics is subject to a high degree of clustering. Moreover, we obtain the implication for the Ohta-Kimura ladder model that if the incremental effect of a single mutation is sufficiently small, then the clustering can be described in terms of nonuniformity of subcell occupation frequency which will appear at a subdivision of size Γ^{-1} such that Γ is large but yet much smaller than $N^{1/2}$, the square root of the population size.

7. Extensions.

Shiga et al. (1986) have extended the ideas of Fleming and Viot to differential operators $\mathcal{L}$ on $M_1(S)$ of the type

$$\mathcal{L}F(\mu) = \tfrac{1}{2} \int\int \big[\mu(dx)\delta_x(dy)\beta(y) - \mu(dx)\mu(dy)\big(\beta(x) + \beta(y) - \langle\mu,\beta\rangle\big)\big]$$
$$\cdot \frac{\delta F(\mu)}{\delta\mu(x)\,\delta\mu(y)} - \int \mu(dx)\big(\beta(x) - \langle\mu,\beta\rangle\big)\frac{\delta F(\mu)}{\delta\mu(x)} + G(\mu)\frac{\delta F(\mu)}{\delta\mu(x)} \tag{7.1}$$

where $\beta(x)$ is a bounded measurable and uniformly positive function on S, and for each $\mu \in M_1(S)$, $G(\mu)$ is a linear functional defined on a dense subset $D(G)$ of $C(S)$ satisfying $G(\mu)1 = 0$; the domain of $\mathcal{L}$ consists of all functions of the form

$$F(\mu) = f\big(\langle\mu,\phi_1\rangle, \ldots, \langle\mu,\phi_d\rangle\big); \qquad f \in C_b^2(\mathbb{R}^d), \ \phi_i \in D(A). \tag{7.2}$$

(Note that if $\beta(x)$ is a constant function, then the operator $\mathcal{L}$ of (7.1) is an operator of Fleming-Viot type introduced in Section 4.) They claim that this operator can be used to model several problems from the areas of population genetics and dynamics.

Fleming and Viot (1979) introduced a Darwinian selection feature into their model by letting a symmetric function $m(x, y)$ denote the fitness coefficient of genotype (x, y), setting

$$B(\mu)(dx) = \Big[\int m(x, y)\,\mu(dy) - \int\int m(z, y)\,\mu(dy)\,\mu(dz)\Big]\mu(dx)$$
$$= \int b(\mu; y)\,Q(\mu : dx \times dy) \tag{7.3}$$

where

$$b(\mu; y) \equiv \int m(y, z)\, \mu(dz), \tag{7.4}$$

$$Q(\mu : dx \times dy) = \mu(dx)\, \delta_x(dy) - \mu(dx)\, \mu(dy), \tag{7.5}$$

and replacing (4.11) by

$$A = D \cdot \Delta + B. \tag{7.6}$$

Application of the Cameron-Martin-Girsanov transformation then yields results about the non-neutral Fleming-Viot model. Since measure-valued processes related by the Cameron-Martin-Girsanov transformation have equivalent probability laws for finite t, any almost-sure sample path properties, such as the Hausdorff-Besicovitch dimension of topological support of the random measures at time t, are inherited.

Until now, we have considered models in which the basic genetic unit is taken to be the gene, the basic numerical quantity is the gene frequency, and the fundamental unit step in evolution is the replacement of one allele by another in the population. However, recent recognition of the gene as a sequence of nucleotides has led mathematical geneticists to consider other models arising from molecular population genetics theory. To some extent, the purely mathematical theory presented so far carries over to the molecular level, with the nucleotide frequency replacing the gene frequency as the primary variable. However, there are clearly aspects in which the molecular theory differs from the classical theory based on gene frequencies. One such aspect is the dynamic nature of the molecular theory: since it is plausible that most nucleotide mutations will lead to sequences not currently existing in the population, mutations should be viewed as leading to new genes rather than to currently or previously existing ones.

Two models which take this into account are the infinitely-many-alleles model of Kimura and Crow (1964) and the infinitely-many-sites model of Kimura (1969). It is worth noting that each of these models has recently been studied via the method of measure-valued diffusion approximations (Ethier and Kurtz (1985), Ethier and Griffiths (1985)). Also, we note that our technique of duality has been applied by Shiga (1981) to infinite dimensional diffusions arising from genetical models and by Donnelly (1984, 1985) to several other problems in population genetics.

Finally, we note that, in fact, nucleotide sequencing has in the last five years entirely superseded electrophoresis as the preferred method for genetic classification.

Acknowledgment. This research was partially supported by the National Science Foundation under grant DMS-83-11972.

References.

[1] D. A. Dawson and K. J. Hochberg, "Wandering random measures in the Fleming-Viot model," *Annals of Probability*, **10**(1982), 554–580.

[2] D. A. Dawson and K. J. Hochberg, "Qualitative behavior of a selectively neutral model," *Theoretical Population Biology*, **23**(1983), 1–18.

[3] D. A. Dawson and K. J. Hochberg, "Function-valued duals for measure-valued processes and applications," *Contemporary Math.*, **41**(1985), 55–69.

[4] D. A. Dawson and K. J. Hochberg, to appear (1986).

[5] D. A. Dawson and T. G. Kurtz, "Application of duality to measure-valued diffusion processes," *Lect. Notes in Control and Information Sci.*, **42**(1982), 91–105.

[6] P. Donnelly, "The transient behaviour of the Moran model in population genetics," *Math. Proc. Camb. Phil. Soc.*, **95**(1984), 345–358.

[7] P. Donnelly, "Dual processes and an invariance result for exchangeable models in population genetics," *J. Math. Biology*, to appear (1985).

[8] P. Donnelly, "Dual processes in population genetics," *Proc. Heidelberg Conf. on Spatial Processes*, to appear (1985).

[9] S. N. Ethier and R. C. Griffiths, "The infinitely-many-sites model as a measure-valued diffusion," to appear (1985).

[10] S. N. Ethier and T.G. Kurtz, "The infinitely-many-alleles model with selection as a measure-valued diffusion," to appear (1985).

[11] W. J. Ewens, *Mathematical Population Genetics*, Springer-Verlag, Berlin-Heidelberg-New York, 1979.

[12] R. A. Fisher, "On the dominance ratio," *Proc. Roy. Soc. Edin.*, **42**(1922), 321–431.

[13] W. H. Fleming and M. Viot, "Some measure-valued population processes," in *Proc. Int. Conf. on Stochastic Analysis*, Academic, New York, 1978, 97–108.

[14] W. H. Fleming and M. Viot, "Some measure-valued Markov processes in population genetics theory," *Indiana Univ. Math. J.*, **28**(1979), 817–843.

[15] K. J. Hochberg, "Microscopic clustering of population processes," *Lect. Notes in Biomathematics*, Vol. 52, Springer-Verlag, 1983, 18–24.

[16] R. Holley and T. Liggett, "Ergodic theorems for weakly interacting systems and the voter model," *Annals of Probability*, **3**(1975), 643–663.

[17] R. Holley and D. W. Stroock, "Dual processes and their aplications to infinite interacting systems," *Advances in Math.*, **32**(1979), 149–174.

[18] R. Holley, D. W. Stroock and D. Williams, "Applications of dual processes to diffusion theory," *Proc. Symp. Pure Math.*, Vol. 31, American Mathematical Society, Providence, 1977, 23–26.

[19] H. Kesten, "The number of distinguishable alleles according to the Ohta-Kimura model of neutral mutation," *J. Math. Biology*, **10**(1980a), 167–187.

[20] H. Kesten, "The number of alleles in electrophoretic experiments," *Theoretical Population Biology*, **18**(1980b), 290–294.

[21] M. Kimura, "Evolutionary rate of the molecular level," *Nature*, **217**(1968), 624–626.

[22] M. Kimura, "The number of heterozygous nucleotide sites maintained in a finite population due to steady flux of mutation," *Genetics*, **61**(1969), 893–903.

[23] M. Kimura and J. F. Crow, "The number of alleles that can be maintained in a finite population," *Genetics*, **49**(1964), 725–738.

[24] M. Kimura and T. Ohta, "Distribution of allelic frequencies in a finite population under stepwise production of neutral alleles," *Proc. Nat. Acad. Sci. U.S.A.*, **72**(1975), 2761–2764.

[25] M. Kimura and T. Ohta, "Stepwise mutation model and distribution of allelic frequencies in a finite population," *Proc. Nat. Acad. Sci. U.S.A.*, **75**(1978), 2868–2872.

[26] J. F. C. Kingman, "Coherent random walks arising in some genetical models," *Proc. Roy. Soc. London A*, **351**(1976), 19–31.

[27] J. F. C. Kingman, *Mathematics of Genetic Diversity*, Society for Industrial and Applied Mathematics, Philadelphia, 1980.

[28] P. A. P. Moran, "Wandering distributions and the electrophoretic profile," *Theoretical Population Biology*, **8**(1975), 318–330.

[29] P. A. P. Moran, "Wandering distributions and the electrophoretic profile, II," *Theoretical Population Biology*, **10**(1976), 145–149.

[30] T. Ohta and M. Kimura, "Simulation studies on electrophoretically detectable genetic variability in a finite population," *Genetics*, **76**(1974), 614–624.

[31] T. Shiga, "Diffusion processes in population genetics," *J. Math. Kyoto Univ.*, **21**(1981), 131–151.

[32] T. Shiga, A. Shimizu and H. Tanaka, "Some measure-valued diffusion processes associated with genetical diffusion models," preprint (1986).

[33] F. Spitzer, "Interaction of Markov processes," *Advances in Math.*, **5**(1970), 246–290.

Kenneth Hochberg received a B.A. summa cum laude from Yeshiva University and an M.S. and Ph.D. from New York University, where he wrote his dissertation at the Courant Institute of Mathematical Sciences under Henry P. McKean, Jr. in 1976. After holding postdoctoral positions at Carleton University and Northwestern University, he came to Case Western Reserve University, where he was privileged to meet and work with department chairman Gail Young. His research interests are in the areas of probability theory and stochastic processes, including applications, especially to the biological sciences. He has recently been granted a Fulbright Scholar Award, which will be spent in Israel.

COMBINATORICS AND APPLIED MATHEMATICS

DANIEL J. KLEITMAN

Department of Mathematics
Massachusetts Institute of Technology
Cambridge, Massachusetts 02139

I have come to speak on the subject of combinatorics, or more generally, discrete mathematics; and applied mathematics.

I welcome the opportunity to speak on this subject for two reasons. First it is one that has personal meaning to me. I began adult life as an applied mathematician (at least in the English sense of the word; as a physicist, actually), became a combinatorialist of a relatively pure kind, and then became an applied combinatorialist as well. Moreover, I have for a long time been a combinatorialist in an applied grouping within my department. Second, it gives me the opportunity, rare for a research oriented mathematician, to pontificate on general mathematical culture, without feeling constrained by circumstances to acquaint the audience with new technical mathematical results.

When I began to prepare for this talk, I thought I would begin, as mathematicians often do, with some definitions. I soon discovered that there were no clear and universally accepted definitions of anything I wanted to discuss. I found myself more and more confused, until I realized that I was trying to use the concept of "mathematics" in several different senses at once.

So let us begin with some definitions of distinct concepts of "mathematics."

What first springs to mind is what we will call *mathematics1*, which is the union of all the results, theorems, researches, facts and folklore that has developed over the centuries in the area of mathematics, which area you may define yourself.

A second, and distinct concept is that which mathematicians concern themselves with and work on; which we will call *mathematics2*. In the best of all possible worlds *mathematics1* and *mathematics2* would be disjoint; since as far as I know, most mathematicians are trying in their work to create new results, interpretations, extensions, and interrelations that are not already part of *mathematics1*. Incidentally, almost all of the burgeoning number of mathematics journals devote almost all of their space to accretions of *mathematics1*, which, if anything, are depletions of *mathematics2*; little if anything is published about what mathematicians are actually trying to do, outside of contract proposals, which are not widely circulated. This greatly contributes to the well known phenomena that most mathematicians have no idea what others in different specialties do. It makes it even less surprising that the general public and even other scientists have no idea at all what mathematicians do.

A third concept, *mathematics3* is the bag of tricks, techniques, ideas, structures, etc., that mathematicians take from the enormous mass of *mathematics1* to use in their efforts on *mathematics2*.

Finally, *mathematics4* consists of the mathematics curriculum which we inflict on our students. If they are at an appropriate level this presumably has significant overlap with *mathematics3*, being still, a subset of *mathematics1* for the most part.

Armed with these distinctions, let us now turn to define the present topic.

In olden days there was no sharp distinction among areas of mathematics, and the notion of "combinatorics" or of "discrete mathematics" had no particular meaning. I have no idea how one could go through *mathematics1* and identify its combinatorial parts. On the other hand there are subjects that are identifiable as parts of *combinatorics2* or *3*.

At first I thought I could define these by reference to subject: they concern finite or discrete problems or methods or those without algebraic structure. But this is just not correct.

If it were, then combinatorics would certainly be disjoint from analysis, which deals with the continuum and with functions defined thereon, among other subjects. But if one has a discrete problem, the standard technique involves defining a generating function, which is typically an analytic function, which you study by analytic techniques, if you can. In this way, the area has as close a relation to analysis as has number theory. Moreover, there are many combinatorial problems that arise in analysis, and there is a large area of overlap in research interests in the two fields. In particular, the classic special functions and various q-analogs have properties that are of interest to both fields, and there is considerable interest in finding combinatorial arguments for questions in this area. There are individuals such as Steve Milne and Dennis Stanton, whose works are of interest in both fields.

If there is no clear distinction between combinatorics and analysis, there is even less between it and other traditional areas of mathematics. Probability, another branch of analysis, has close relations with combinatorics; there is an area of topology called combinatorial topology, and interrelations between the fields are not dead. There are many kinds of combinatorial geometry, both relating to finite geometries and to combinatorial problems in geometrical contexts. There are logical questions that closely relate to finite combinatorics, and an area called infinite combinatorics that relates to both. There is much common ground between group theory and combinatorics. And so on, and so on. There are even relations between combinatorics and algebraic geometry.

All of which conveys no idea whatever of what the subject is and why it is considered a separate branch of mathematics.

Perhaps the answers to these questions lie in the history of mathematics in this century. In the middle of the century there developed the notion that there is a common general structure to mathematics, and an attempt was made to recast it all in this general mold; the names Dieudonné and Bourbaki among others are associated with this attempt. There is certainly a beauty and grandeur to this concept and effort.

In this context, combinatorial questions and combinatorial arguments were those that did not fit in well with this general structure. They represented the special as opposed to the general, the detailed case by case argument as opposed to the sweeping theorem-ex-machina, the ugly as opposed to the beautiful among aspects of mathematics.

Somehow worrying about the special or combinatorial problem got to be considered a low brow occupation. Graph theory, for example, was commonly referred to as "the slums of topology."

I discussed these questions with a young colleague, and he made two interesting points. He recalls, first, the reaction of some of his teachers when he told them he was interested in combinatorics. They looked at him exactly as if he had informed them that he had decided to become a shoe salesman.

He also noted one of the flaws of the quest for generality. In studying vector spaces, of polynomials for example, the general approach avoided the use of bases as much as possible, which was entirely. It had very little to say about such spaces, in consequence. It so happens, however that there are a number of different bases of polynomial spaces, and their interrelations have a tremendous amount of fascinating structure, that relates to many branches of mathematics, and which have been the subject of much recent interesting combinatorial work. This entire subject with its immensely rich mathematics, vanishes if one refuses to introduce bases because they represent unnecessary specification.

To digress a little, I recall something even sillier from my own education. I recall purchasing an old physics text that was quite nice, but the author had a passionate hatred of vector notation. This was so great that he refused to write any vector or tensor equations: every equation was written out component by component; which was probably great for him if he was paid by the symbol. Of the two passions, against vectors or against bases, I suppose that against vectors is the more ludicrous; fortunately we can avoid either one.

To summarize this somewhat diffuse argument, combinatorics came to represent those areas of mathematics in which one got very little mileage out of the general approach, and those ugly special problems in every area of mathematics which arise upon introduction of special structure. In a sense, from the generalist point of view it was the epitome of ugliness in pure mathematics.

Of course it is the work of mathematicians to bring order, structure and beauty to seemingly unharmonious and ugly situations. "Combinatorics" should therefore have been seen as a mine pit from which future mathematics would be extracted.

Instead, because of prejudice against the area, it was relatively neglected. As a result it was, for a time, and may still be, an unusually fruitful one for mathematical development.

There is much more controversy over what applied mathematics is than over what combinatorics is.

The obvious definition would be: it is the application of *mathematics1* or *mathematics3* to other subjects.

But this is plain wrong. Is an accountant who adds up a column of figures an applied mathematician? Is a physicist or economist who solves a differential equation or a system of equations by standard methods?

This sort of work can be performed by anyone technically versed in mathematics. An applied mathematician may occasionally do this, but he or she would not consider it applied mathematics.

Rather, one view of applied mathematics is that it is the kind of *mathematics2* that arises in the course of attempting to apply *mathematics3* as above.

Applied mathematicians often try to solve problems in other fields using mathematics. When the problem turns out to develop ramifications that are not standard, the technician is made ill; the applied mathematician instead is delighted. He or she has found a new problem and can develop the *mathematics1* needed to attack it.

One problem with this area is that successful work often goes into *sciences3* or *engineering3* rather than *mathematics3*. Most mathematicians never have any contact at all with it, and are unaware of it.

But we have not yet adequately defined applied mathematics. There is no doubt that work on formulating and attempting to solve a new problem that arises in some area outside of mathematics is applied mathematics. But there are still two issues unresolved that are germane to our subject.

One is, can applications of mathematics within mathematics be applied? How about applications of a method from mathematical logic in analysis; or of algebraic geometry to combinatorics? I see no difference in spirit between such applications and those to physics or chemistry.

A more difficult question is: where do you draw the line in distinguishing between the applied work, and similar non-applied work?

Suppose, for example, A discovers and formulates an applied problem X, but does not succeed in solving it completely. B works on this problem with no direct contact with the application. An obstacle to solution of X is noted by C, and D works on methods to overcome this obstacle. E is an expert in the overall area of all this work, and suggests ideas to D on how to handle his problem. Now which of these people are doing applied mathematics?

Some would call only A the applied mathematician; others would extend that role to B, to E, to all of them, and possibly to other combinations of them as well. There are at least two schools of thought on this matter, and their disagreement is not always rational and polite.

There is another definition of applied mathematics that is also relevant to us, and that is Hardy's.

Hardy was a fine mathematician, and, I am told, a very nice person, but he had a peculiar view of applied mathematics. We have noted that from a certain general point of view combinatorics represented everything unpleasant about mathematics. To Hardy, applied mathematics represented everything hateful and revolting, and, yes, evil, about mathematics.

He wrote a book, unfortunately an influential one, which consisted of little more than a diatribe against applied mathematics. It was, he said, ugly, unpleasant, unmathematical. Nothing applied could be mathematically interesting or aesthetic. He thanked his creator that number theory, his own favorite interest, was and would always remain utterly pure, unsullied by the slightest application.

I have no ready explanation for this attitude. Some people, I have discovered, are put off, when they hear about work that does not interest or involve themselves. They feel that its very existence implies that there is something wrong with themselves or with the work, for otherwise they would be interested in it. Perhaps Hardy reacted this way to applied mathematics. He found that his enjoyment of his own accomplishments, his mathematical aesthetics, did not have positive correlation with the utility of his work, and this upset him; so he renounced all application.

Of course one should realize that application in those days was grueling and tedious work. It had two major problems.

First, even such simple tasks as calculating a square root to five places were annoying and time consuming, or you used interpolation on a table of logarithms. Solving a recursion, or a differential equation or doing an integral that was not quite standard was a horrible and very time consuming chore. One had to be extraordinarily lucky if the given problem was one of the special few with calculable solutions.

Second, the non-mathematician who brought you the applied problem, then as now, usually had it somewhat wrong. If you were not familiar with the background area, often as not you worked like a dog at the dullest sort of numerical drudgery to solve it, only to be told, when you brought him your answer, that he was sorry that he had it wrong, would you please solve a slightly different problem.

Perhaps Hardy had some bad experience of this kind, that drove him to his mad dog views.

It should perhaps be noted, that applied mathematics has gotten much pleasanter in recent years. The computer is capable of banishing all of the traditional drudgery. For a while the drudgery was replaced with the equally repellant need to interact with computer centers and programmers or card punch machines. We are now the first generation which can actually perform even the very worst computations imagined by Hardy, and have fun doing it, on our kid's toy computer. It still pays to know about the area of application if you want to do applied mathematics, and this is still a serious hurdle that limits entry into the field.

In one sense, of course, Hardy was wrong. Numbers, at least natural numbers, are a common heritage of mankind, and not really a part of mathematics. Thus number theory, which concerns properties of such things, is really a species of applied mathematics where the application is to the properties of these entities. His insistence that number theory is the apotheosis of purity in mathematics is therefore questionable from the start.

So much for definitions; let us turn to our subject, the interaction of combinatorics with applied mathematics.

If we choose appropriately broad definitions, combinatorics can be considered a kind of applied mathematics, where the applications are often to mathematics itself.

The effect of the computer has greatly enlarged the powers of the applied mathematician, and much of that has come from discrete mathematics. After all, the computers that we use are digital, and we use them for mathematical chores by making the latter into discrete problems and using finitary techniques to solve them. Thus discrete approximations have become the central tool in the numerical analysis which has revolutionized the power of applied mathematics.

I don't want to minimize either of these relations between combinatorics and applied mathematics, but to the extent that the subjects have come to be one and the same, one can hardly speak of their interaction.

Instead I will point your attention at four examples of situations in which discrete problems and applications interact in such a way as to change our notions about applications in important ways.

In examining the potential for applying computers to mathematical and other tasks, theoreticians were led to study and attempt to rate the comparative difficulty of problems. This subject, sometimes called "complexity theory," has both theoretic and practical aspects.

One important open question in the area is: are there classes of problems such that it is always easy to verify a solution to a problem in the class, but always hard to solve it if you don't know a solution.

It is unknown whether such a class exists, but we suspect that it does. If it does, and we can locate such a class of problems, there are important practical applications that flow immediately from it. In particular, the problems can be used for generating secret codes. The very difficulty of solving the problems can be converted into a corresponding difficulty in breaking the code. And by the way, secret coding is now an extremely important subject, relevant to computer communications, computerized banking, and national security.

The implication of all this is that any area of mathematics that possesses or may possess "one way hard problems" as just described, is directly or potentially applicable.

Now it so happens that the very first area in which a potential problem class was suggested, to be used in this way, was number theory; the problem that of factoring huge numbers that are the product of two primes.

An early paper on this subject received thousands of reprint requests including those from governments all over the world. Suddenly, any results or ideas in number theory that might lead to new more efficient factoring methods became of great practical interest, and there has been quite a bit of interesting work in this area. A branch of our government at one time toyed with the idea of classifying research in number theory, and there are now some restrictions on publication in this area.

One consequence of all this is the utter collapse of Hardy's view of applied mathematics.

Furthermore, there is no branch of pure mathematics that can safely proclaim that it has no "one way hard problems," and is therefore immune from applicability. In fact, the quest for problems of this kind is a new applied problem area within every branch of mathematics.

To me, this development is extremely important and has very important implications for government funding.

The atomic bomb proved that the most esoteric physics of its day had enormous practical application. The physics community has parlayed that lesson and subsequent developments to justify support of multibillion dollar efforts in narrow research areas far less likely to have application than the average field of pure mathematics.

The potential application of pure mathematics to secret coding has a similar potential for the funding of mathematics.

And of course it necessitates a complete rethinking of the definition of applied mathematics.

My second example is Karmarkar's algorithm for linear programming. Linear programming is the problem of finding a point in a region in n-space bounded by hyperplanes, that has largest component in some direction. The standard (simplex) algorithm is wonderfully if somewhat mysteriously efficient, and it proceeds by wandering from corner to adjacent corner of the region, always representing the problem in terms of a basis of variables that vanish at the corner one is on. Karmarkar made use of ideas of projective geometry and a projectively invariant representation of the problem to develop an algorithm that is not only provably efficient for the problem, but seems to be practically even better than the simplex algorithm for many large problems. It has the potential of increasing the size of solvable LP problems by an order of magnitude.

Here we see projective geometry finding application and practical application in discrete optimization.

The third example involves error correcting codes. This widely applied subject concerns codes whose words are vectors in a space over a finite field, and sufficiently far way from one another that one can still recognize a code word even if a number of its entries are wrong. In this field one can prove that there are very efficient codes of this kind by probabilistic arguments, but practical codes, for which there are convenient ways to construct the codes and to decode in the presence of errors, are not so efficient. In fact there is the called "Varshamov bound," which has described an upper limit to the efficiency of such codes. In the last few years however, by using relatively new developments in algebraic geometry, several young Russians were able to develop codes, whose words are integral points on certain obscure varieties, that beat this bound.

These codes are not now actually of practical interest; but here we have application of reasonably current work in algebraic geometry.

The fourth example shows a complicated interaction: an application (1982) by Steele and Yao of a theorem of Milnor from the 1960's to finding lower bounds on the complexity of algorithms in computer science used a method that has led to a number of applications to pure combinatorical enumeration problems.

Suppose a region in n dimensions is defined by k equations each of degree at most m. Milnor obtained an upper bound on the number of connected components of this region, in terms of these parameters. Steele and Yao related connected components with permutations and were able to use this result to show that at least a certain amount of effort is required to solve any member of a general class of sorting problems.

Goodman and Pollack, Ben-Or and Alon, Frankl and Rodl have made use of this and related results from real algebraic geometry to get new bounds on, for example, the number of convex d-polytopes on n points.

This example is particularly appropriate for this meeting. Steele and Yao in their paper [1] recommend that readers unfamiliar with Milnor's result and its context read a text by Gail Young which they commend as a clear and lucid introduction to the subject.

What are we to conclude from all this?

Many of the notions about applied mathematics prevalent in the mathematics community have been left behind by events, and in particular by the implications of the development of computers.

There has not only been extensive recent development in discrete applied mathematics but there have been and will be important applications of what has until now been considered pure mathematics in this area.

There are implications here about the mathematical depth and breadth of the education of applied mathematicians that we ignore at our peril. The more exposure future applied mathematicians get to pure mathematics the better.

The implications of the present situation for the funding of pure mathematics have not been adequately developed and put across by pure or applied mathematicians.

References.

[1] J. M. Steele and A. C. Yao, "Lower bounds for algebraic decision trees," *Journal of Algorithms*, **3**(1982), 1–8.

Daniel Kleitman received his bachelor's degree from Cornell University in 1954, and a master's (1955) and Ph.D. (1958) in theoretical physics from Harvard, where he was a student of Julian Schwinger and Roy Glauber. He was a Postdoctoral Fellow in Copenhagen and Harvard, and taught physics until 1966 at Brandeis University, joining the Mathematics Department at MIT in that year, where he has remained ever since. He has served as Chairman of the Applied Mathematics Committee in that Department and as Department Head. His main interests are in combinatorial enumeration and extremal problems; he has also consulted on a wide variety of subjects, including nuclear safety issues, network design, and applied statistics. He is also an antiques dealer.

APPLIED LOGIC

A. NERODE

Department of Mathematics
Cornell University
Ithaca, New York 14853

INTRODUCTION.

Throughout the centuries the great themes of pure mathematics, which were conceived without thought of usefulness, have been transformed to essential tools for scientific understanding. This lecture is devoted to the theme that this transformation is now happening to mathematical logic, and that a subject of applied logic is emerging akin in its range and power to classical applied mathematics. This adds further weight to the argument that the investment of intellectual and material capital in mathematical research pays a rich dividend.

Mathematical logic is a branch of mathematics originating in the study of philosophy and foundations of mathematics. For two thousand years, from Aristotle to the advent of digital computers, logic had no significant applications. Now there are manifold uses of ideas in mathematical logic which were thought recondite only a few years ago. These applications are currently to mathematics, linguistics, and computer science. Future applications are likely to be in those other scientific and engineering areas which need massive automated logical inference. We confine ourselves to computer science applications. We do not address the specialist in either logic or computer science, but simply give an uneven impressionistic account and a long list of references to indicate the scope of the applications of logic in computer science. Neither the account nor the list of references exhausts important applications or important contributors. For organizational convenience we separate applications into two groups on the basis of historical connections with two of Gödel's achievements, the completeness theorem (Gödel [1930]) and the incompleteness theorem (Gödel [1931]). The separation is a little artificial, and neither group of applications can be fully understood without the other.

An historical approach brings to the surface the debts that computer science owes logic. On the other side, logic owes computer science a great debt for presenting it

127

with a multitude of interesting new challenges. The challenge posed by computer science may well provide the kind of stimulus to logic that physics has provided for analysis since Newton.

PART 1. APPLICATIONS ASSOCIATED WITH COMPLETENESS.

1.1. Propositional Logic and Its Automation.

The propositional logic of 'and' (denoted by '$\cap$', also called conjunction), 'or' (denoted by '$\cup$', also called disjunction), and 'not' (denoted by '$\neg$', also called negation) goes back to pre-Socratic times. This is the only part of logic that everyone knows, and that everyone knows is applicable mathematics (switching circuits, circuit design). Propositional logic and its applications are assumed as known by the readers of this lecture. Propositional logic was discovered anew by Leibniz, who dreamed of a calculus of reasoning, and was rediscovered again by many early nineteenth century authors. Propositional logic was developed by Boole [1854] in algebraic form in his famous *Laws of Thought*. He was interested in how to identify the logical truths (tautologies, statements true under all possible truth assignments). His procedures were tedious. The question as to how efficient these procedures can be made is Cook's now-famous $P = NP$ problem. Propositional logic statements are built up from simplest statements (called atomic statements) using the logical connectives '$\cap$', '$\cup$', '$\neg$'. Boole approached the determination of whether a statement is a tautology using his "Law of Expansion". This amounts to reducing a propositional logic statement to disjunctive normal form (a disjunction of conjunctions of atomic statements and their negations). E. L. Post [1920] gave a formal system of rules of proof for propositional logic and proved these rules were complete; that is, he proved that any logical truth ϕ of propositional logic has a proof from his axioms by the deduction rules of his system. R. A. Robinson [1965], improving on ideas of Putnam, Davis, and Wang (see Siekmann-Wrightson [1983]), developed a proof rule for propositional calculus called the resolution rule. This is a single rule of proof (with no axioms) which suffices for proving all tautologies of propositional calculus. It is suitable for automation of propositional logic by computer programs. To prove a statement ϕ using the resolution rule, we try to show (for a contradiction) that there is no truth assignment making $\neg\phi$ true. First reduce $\neg\phi$ to conjunctive normal form, a conjunction of disjunctions ς_j of atomic statements and their negations. These disjunctions are called clauses, the clauses obtained from ϕ. The atomic statements and their negations occurring in the disjunctions are called literals. Our inference rule deduces clauses from clauses. To show that $\neg\phi$ cannot be made true, we start with the clauses obtained from ϕ as premises, and use the following rule as the sole rule of inference: if two (parent) clauses share a propositional letter P, with literal $\neg P$ in one clause and literal P in the other, then infer the clause obtained by omitting these occurrences of P from both clauses and taking the disjunction of the remaining literals in both clauses. This is called resolving P against $\neg P$. A contradiction occurs if the empty clause is reached. If we get a contradiction, we have proved ϕ by a resolution proof. Improvements on this rule (such as linear resolution) have been an active area of research (see Loveland [1978]). Each improvement of the resolution rule is designed to make automated reasoning more efficient for some class of statements or more efficient on some kind of machine.

1.2. First Order Logic and Its Automation.

Syllogistic. The logic of the quantifiers 'for all' (denoted by '$\forall$'), and 'there exists' (denoted by '$\exists$'), was developed by Aristotle as syllogistic in the Golden Age of ancient Greece. The purpose of syllogistic was to establish algorithmic checks for the validity of philosophical arguments. Syllogistic was the earliest "proof-checker", and the precursor of automated reasoning with quantifiers. This is a logic dealing with properties which apply to objects in a domain. These are objects such as Gail Young, and properties such as 'x is mellow', or 'x is wise'. Syllogistic deals with quantifiers over such objects. This logic contains rules for deducing conclusions from premises. For example,

from '$(\forall x)(x$ is wise implies x is mellow)'

and '(Gail Young is wise)',

deduce '(Gail Young is mellow)'.

Syllogistic dealt conveniently only with properties of objects, not relations between objects, although users had some awkward devices for reducing relations to properties ('x is the father of y' reduced to "x has the property 'is the father of y'"). As a result, syllogistic's capabilities fell short of the logical needs of the deductive mathematics being developed by Aristotle's contemporaries. Euclid's Elements, to cite an example, was based on such relations between objects as 'line x is incident to plane y', rather than just properties of objects. Convenient representation of relations is essential to carrying out deductive mathematics. Thus neither the mathematical axioms nor the relational logic of Euclid were formalized in ancient times. Instead, for two thousand years these methods were learned by the study and imitation of Euclid and then used by mathematicians as a model to develop the rest of mathematics. Deductive mathematics was identified with geometry, mathematicians were often referred to as geometers, and geometry was taken to be the foundation of mathematics. The developments following Newton in analysis outstripped this foundation entirely. After Weierstrass's arithmetization of analysis laid a firm foundation for analysis in the 1850's, late nineteenth century mathematicians such as Peano, Schröder, and Frege generalized syllogistic by writing out the logic of relations and quantifiers. They were able to express all known mathematical propositions and their proofs. Frege [1879] was the first to publish a definition and exact formal rules of proof for first order logic. We give a description of first order logic so that our later discussion will make some sense to those with no logical training, and apologize to those who are well acquainted with first order logic.

First Order Logic. The primitive vocabulary of first order logic consists of the propositional connectives and quantifiers ($\forall$, $\exists$, $\cap$, $\cup$, $\neg$), n-ary relation symbols R (intended to name n-ary relations on a domain of objects), constants (intended to name objects in a domain), n-ary function symbols (intended to name n-ary functions on a domain), a list of variables (intended to range over the domain), together with commas and parentheses. Atomic formulas are expressions in this vocabulary of the form $R(t_1, \ldots, t_n)$, where $t_1, \ldots, t_n$ are terms built up from constants and variables by composition using function symbols. Formulas of first order logic are defined by the following definition.

(1) Atomic formulas are formulas.

(2) If α and β are formulas, so are $(\alpha \cap \beta)$, $(\alpha \cup \beta)$, $(\neg \alpha)$.

(3) If v is a variable, and ϕ is a formula, then $(\forall v)(\phi)$ and $(\exists v)(\phi)$ are formulas.

(4) Nothing else is a formula.

An occurrence of a variable v in a formula ϕ is called free if that occurrence does not occur within a subformula of ϕ which begins with a quantifier $(\forall v)$ or $(\exists v)$, otherwise that occurrence is called bound. The simplest statements are the atomic statements, the statements involving no logical operations. These are the statements of the form $R(t_1, \ldots, t_n)$, where $t_1, \ldots, t_n$ are terms built up from function symbols and constants (but no variables) by composition.

The logic built up from such formulas is called first order because only quantification over objects of the domain occurs (first order quantification). First order logic does not allow quantification over subsets of the domain, or over subsets of subsets of the domain, as does higher order logic.

First order logic has been a main theme in mathematical logic research since the 1920's. It is both quite expressive (Zermelo-Fraenkel set theory, encompassing most of mathematics, is a theory in first order logic expressed by a countable sequence of axioms), and is also susceptible to deep proof-theoretic and algebraic investigations (proof theory and model theory). Higher order logic is more expressive than first order logic, but has not proved as amenable to algebraic or proof-theoretic investigations.

The semantics of first order logic makes precise the notion of interpreting a formula in different ways in different domains of objects. An interpretation (now called a relational system) is determined by specifying the following: a non-empty set as domain; a relation on the domain for each relation symbol to denote; an n-ary function on the domain for each n-ary function symbol to denote; and an object in the domain for each constant to denote. Thus a statement may hold in one interpretation and fail in another, as in the reals 'every number has a square root' fails, but in the complex numbers this same statement is true.

The Skolem-Löwenheim Theorem. Out of Schröder's massive extension of Boole's algebra of logic in the 1890's came an extensive development of domains and interpretations, and the first significant theorem about first order logic. This was Löwenheim's proof [1915] that if a first order statement is true in some domain, then it is true in some countable domain. This is now known as the Skolem-Löwenheim theorem (see Skolem [1920]). We describe their proof because it contains ideas which are precursors to those in automated reasoning.

Suppose R is a logical formula without quantifiers and ϕ is the statement below.

$$(\forall x_1) \ldots (\forall x_n)(\exists y)\big(R(x_1, \ldots, x_n, y)\big).$$

Suppose there is at least one domain D making ϕ true, with appropriate assignments of meaning to relation symbols, function symbols, and constants. To settle the question whether there is then a countable domain with the same property, we construct a slightly modified statement ϕ^S in which the y asserted to exist corresponding to any $x_1, \ldots, x_n$ is viewed as a function of $x_1, \ldots, x_n$. That is to say, we introduce a new function symbol $f(x_1, \ldots, x_n)$. Observe that there is a domain D making ϕ true if and only if there is a domain D' making the ϕ^S below true.

$$(\forall x_1) \ldots (\forall x_n)\big(R(x_1, \ldots, x_n, f(x_1, \ldots, x_n))\big).$$

Let t be any element of D'. Let D'' be the smallest subset of the domain D' which contains t and any other elements named by constants occurring in R and is closed under all functions (including f) denoted by function symbols occurring in ϕ^S. Since D'' is generated from a finite number of elements by a finite number of functions, D'' is countable. Obviously it is a domain in which ϕ^S is true. So ϕ is true in D'' too. Thus if ϕ is true in any domain, it is true in a countable domain, and the Skolen-Löwenheim theorem is proved for ϕ. To generalize this proof of the Skolem-Löwenhein theorem to arbitrary statements ϕ of first order logic, drive all quantifiers of ϕ to the front by standard logic rules to get the so-called prenex form, so that the rest of the formula after the initial string of quantifiers is a quantifier free formula R. Eliminate all existential quantifiers by introduction of function symbols repeatedly as was done once above, obtaining ϕ^S with only universal quantifiers, reducing the problem to the case already treated.

This argument is easily modified to show that if a countable sequence of statements is simultaneously true in some domain, then that sequence is simultaneously true in a countable domain. That this says something mathematically interesting is clear when this is applied to Zermelo-Fraenkel set theory ZF: if ZF is true in some domain, then ZF is true in a countable domain. Yet ZF proves the existence of uncountable sets. This is called Skolem's paradox (it is not a paradox).

An Herbrand Proof. Herbrand [1930] recast the Skolem-Löwenheim argument syntactically. Introduce a new constant c to denote the element t used above. Let H be the set of all terms built up by composition from c and the constants occurring in ϕ^S using all the function symbols occurring in ϕ^S —e.g., terms like

$$f\big(c, f(c, \ldots, f(c, \ldots, c)), c, \ldots, c\big).$$

If there is a domain making ϕ true, then there is a domain D'' as above making ϕ^S true such that every element of D'' is named by a term in H. So we might as well identify D'' with H. (Warning: for applications to mathematics, the equality relation requires special treatment as an equivalence relation with congruence properties.)

In H each relation symbol denotes a relation among terms in H. But now we don't have to worry about the function symbols and constants, each function symbol f acts naturally on terms $t_1, \ldots, t_n$ to produce the term $f(t_1, \ldots, t_n)$ as value and each constant names itself. Nowadays H is called an Herbrand universe for ϕ^S. Since ϕ^S is the universal quantification of R, making ϕ^S true in H by interpreting relation symbols as relations on H reduces to making simultaneously true all the countably many quantifier-free statements R_i which result from simultaneously substituting terms from H for all the variables in R in all possible ways. These statements R_i, called ground instances of ϕ^S, are propositional combinations of atomic statements. They are therefore in the propositional logic based on the atomic statements occurring in the R_i. So Herbrand reduced the question of whether there is at least one domain with ϕ true to an equivalent propositional logic question: is there an assignment of truth or falsity to the atomic statements occurring within the (infinitely many ground instances) R_i which makes all the R_i simultaneously true?

Gödel [1930] was the first to publish a proof of the completeness of first order logic. This is the theorem that every logically true first order statement (that is, every statement true in all non-empty domains) has a proof by a fixed set of rules of proof. The essential content of this theorem is that there is an algorithm for listing all logical truths. Gödel's proof uses the Skolem-Löwenheim idea plus a compactness argument. Herbrand was a constructivist and did not accept the compactness argument required to convert his proofs into a proof of the completeness theorem. Gödel, always a Platonist, had no such compunctions, and has sole credit for stating and proving the completeness theorem. On the other hand, Herbrand's ideas have been the influential ones for the automation of reasoning. We give the rest of the argument for the completeness of first order logic based on Herbrand's ideas. Suppose ϕ is a statement of first order logic and ϕ is logically true. We find a proof of ϕ as follows. Observe that $(\neg\phi)$ is true in no non-empty domain. Form $(\neg\phi)^S$ as the universal quantification of a quantifier free formula R. Look at the Herbrand universe H for $(\neg\phi)^S$, and form the R_i for $(\neg\phi)^S$ by substituting terms from H for all the variables in R in all possible ways. Since $(\neg\phi)^S$ is true in no non-empty domain, there is no propositional truth valuation of the atomic statements in the R_i making all the R_i simultaneously true.

At this point we need the compactness theorem of propositional calculus. This says that for any set A of propositions, there is a truth assignment making all of A simultaneously true if and only if for each finite subset A' of A, there is a truth assignment making that finite subset A' true. This is simply the compactness of the space of all assignments of truth and falsity to the atomic statements occurring inside the R_i, a space which is a product of two-element discrete spaces.

By the compactness theorem for the propositional calculus, there is an n such that $\sigma_n = \neg(R_1 \cup \cdots \cup R_n)$ is a tautology. Now with this we have a procedure for algorithmically listing all logically true statements of first order logic. List all statements ϕ of first order logic. For each ϕ list its sequence $\{R_i\}$ of ground instances and test for each $\sigma_n = \neg(R_1 \cup \cdots \cup R_n)$ whether σ_n is a tautology. Record ϕ in the list of logical truths as soon as one of its σ_n is identified as a tautology.

Unification. Herbrand observed that in going from the variables in R to the statements with no variables R_i, a lot of duplication of effort in testing the σ_n as tautologies is inadvertently introduced. Some of the cases tested are basically special cases of others. He defined a notion of "most general" cases and stated a lemma showing that only these need to be looked at. This lemma contains what is now called the unification theorem for first order logic. It has spawned a vast literature and many applications because of its power as a pattern matcher. There are algorithms implementing the unification theorem which run in linear time (Paterson and Wegman [1976], Martelli and Montanari [1976]). But most implementations of unification avoid implementing a full algorithm for unification for the sake of yet greater efficiency.

Here is a rough introduction to unification. (See Chang and Lee [1973] or Loveland [1976] for the use of unification in first order logic, see Huet and Oppen [1980] for the literature.) By a substitution τ we mean a mapping τ of variables to terms (possibly containing variables). Let τt be the result of substituting τv for v in t for each variable v that occurs in term t, and similarly define τR for quantifier free formulas. Say that terms s, t unify by unifier τ if there is a substitution τ with $\tau s = \tau t$. A most general unifier of s and t is a unifier σ such that for every other unifier κ for s, t there is a unifier λ with $\lambda(\sigma t) = \kappa t$ —i.e., every unifier factors through the most general unifier. Make a similar definition for quantifier free formulas. Herbrand's unification algorithm tells whether or not terms can be unified, and produces a most general unifier if they can be unified. We omit discussion of the algorithm itself. Unification in a broader more algebraic context has become a major area of research in automated deduction, especially in connection with rewrite rules (Huet and Oppen [1980], Siekmann [1984]).

Resolution Plus Unification. R. A. Robinson [1965] combined his resolution method for propositional calculus discussed above with Herbrand's unification method for terms to give a single rule of proof for predicate logic. First, as above, compute $(\neg\phi)^S$ and its quantifier-free R, and reduce R to a conjunction of disjunctions of atomic formulas. These disjunctions are called clauses. The idea is (again as in propositional logic) to prove ϕ by showing that a single rule of inference leads from these clauses to the empty clause. The resolution rule applies to two clauses C, D, for which there exist disjuncts $P_1, \ldots, P_n$ of one and $\neg P_{n+1}, \ldots, \neg P_{n+m}$ of the other and a most general unifier τ for $P_1, \ldots, P_{n+m}$. The rule resolves τP_1 against $\tau(\neg P_{n+1})$, as in propositional logic. This method has been refined in many ways, to be more efficient for certain classes of statements.

1.3. Horn Clause Logic and PROLOG.

The development of the PROLOG programming language in the early 1970's by Colmerauer et al. [1972] and Roussel [1975] was a major event in automated deduction, capping a decade of work. Kowalski ([1974a], [1974b]) observed that PROLOG is best thought of as an implementation of resolution for Horn clause logic, and therefore as a special part of automated first order logic.

Horn clause logic is the logic of the universal quantifications of formulas

$$(P_1 \cap \cdots \cap P_n \to Q),$$

where $P_1, \ldots, P_n, Q$ are atomic formulas. (This is written $(Q: -P_1, \ldots, P_n)$ in PRO-LOG as an artifact of a notation for clauses in resolution theorem proving.) The proof method usually implemented is Robinson's resolution method with simplications for Horn clauses (SLD resolution, Lloyd [1984]). If PROLOG is implemented with a complete theorem prover for Horn clauses (by so-called "breadth first" search of the tree of possible deductions), then PROLOG can properly be regarded as a language in which the logical description of a program as Horn clauses is the program itself (or to paraphrase Kowalski, the specification of the problem in logic is already executable code.) That is, if one implements PROLOG according to logic theory, order of execution is not mentioned explicitly in the program and the interpreter or compiler determines the order of execution.

For efficiency, current commercially vended systems implement only part of the full proof procedure of Horn sentence logic (only depth first search), and only part of the unification algorithm is implemented (the occurs check of the unification algorithm is omitted). This causes major problems both in logical interpretation of programs and also in writing programs that behave as intended. Much of the beauty of logic specification as program is lost.

PROLOG interpreters and compilers have added separate control constructs to the logic (such as 'cut') for handling flow of control. These are thus far less flexible than the control constructs of such languages as PASCAL and LISP. A major current concern is what kind of control constructs should be added to PROLOG, and how the Horn logic should be extended and interpreted.

Logic as programming is a very active field. For the theory behind PROLOG, see Lloyd [1984]. For how PROLOG is implemented, see Hogger [1985], and Campbell [1984]. Horn clause logic has been investigated extensively for parallel execution in the PARLOG of Clark and Gregory [1982], and the concurrent PROLOG of Shapiro (1983). The idea for parallel execution is simply that clausal form breaks up computations into those within clauses and those between clauses, and these computations can be performed in parallel. The original plan for the Japanese fifth generation computer initiative emphasized this advantage as a reason for developing parallel PROLOG machines.

PROLOG is an easy language to learn, much easier than LISP. It is also an easy language for writing symbolic manipulation programs. Sometimes mathematicians want to find counterexamples to conjectures by using symbol manipulation, but the desired counterexamples are too large to construct with pencil and paper. Often writing down the program for constructing the counterexample in PROLOG is easy, because the way PROLOG handles recursive definitions is the way mathematicians handle them. A complete implementation of a PROLOG compiler and interpreter, in which the occurs check can be turned on or off, and breadth first search as well as depth first search is an option, would be a valuable exploratory tool for mathematicians.

1.4. Logic and Knowledge Bases.

Expert and Rule Based Systems. A very successful application of automated reasoning has been the creation of expert and rule based systems for government agencies and industry. Expert systems encode a knowledge base obtained from someone who is thought to know what he or she is doing. Expert systems use automated logical inference to draw conclusions from this knowledge base and give advice.

The three most important problems in expert systems are the problems of knowledge acquisition, knowledge representation, and automated inference. The knowledge acquisition problem is the problem of acquiring general rules from those special instances known to the expert. This is a problem of inductive inference. What is wanted is algorithmic generation of general rules from known instances, rules which then may be applied to yet unexperienced instances. If we perform knowledge acquisition using an expert, we have the advantage of being able to query the expert (after the generation of new rules) about the suitability of these rules. We expose the expert to new instances of the application of the rules, to see whether the conclusions drawn by the rules are the same as the ones the expert would have drawn. This presents interesting questions as to what kinds of logical systems should be used for knowledge acquisition. There are interesting implementations of ideas of this kind in the two expert system writing programs, RULEMASTER and TIMM. The problem of knowledge representation is the problem of choosing a logical language rich enough to cover the intended applications, but simple enough for efficient implementation. These languages can be quite simple (semantic nets, or attribute-value systems) or quite powerful (first order logic and its extensions). See Brachman and Levesque [1985], and the last section of this lecture.

Many implementations of expert and rule based systems are written directly in LISP, many are written in PROLOG, some are written in C, some are written in PASCAL, some are written in production system languages such as OPS5. More and more are written in specialized "expert system shells". All share the use of efficient automated logical inference. To get an overview of the current uses, see Goodal [1985] or Waterman [1985]. These uses include automated advisors for maintenance of machines, analysis of organic molecules, oil prospecting, etc.

Relational Databases. Codd ([1970], [1972]) viewed stored rectangular arrays as binary relations and applied the operations of the logic of relations (calculus of relations) to manipulate such arrays. Commercial relational databases exploit this. Stored knowledge is retrieved by querying the database for all database entries satisfying a logical formula. Many logical systems which express such queries have been used for this purpose. Such computer languages as PROLOG, KRC, FQL, HOPE, and such database languages as QURL, SQL, QBE, ASTRID are modelled conceptually either on the lambda calculus or on many-sorted first order logic. Lambda calculus will be treated in Part 2. A good source of information on logic applied to databases is Gallaire-Minker [1978, 1980]. See also Gray [1984], Jacobs [1985], Kowalski [1981], van Emden [1978], Dahl [1981], Gallaire [1981]). The relational database models are not the only models used for databases. There are also hierarchical and network models, for which development of other logics is appropriate. There is a growing field of research into extensions of first order logic to serve as query languages. This has given rise to some penetrating theoretical questions about the scope of such logics. (See Aho-Ullman [1979], Immerman [1982], Vardi [1982], Gurevich [1984], Gurevich-Shelah [1984]).

Many-Sorted Logic. Many-sorted first order logic was developed in the 1930's to extend first order logic to cover in a natural way algebraic structures which involve more than one domain. This includes such common structures as vector spaces over a field and chain complexes of groups. Many-sorted first order logic deals with relations between elements of different domains. Each of its quantifiers range over exactly one of these domains. Its functions map a cartesian product of domains to a domain. Its model and proof theory are not much different from those of ordinary first order logic. Indeed, many-sorted logic has an easy translation into ordinary first order logic by the simple procedure of adding one unary relation symbol to denote each domain, with axioms to make the domains disjoint. When the proof procedures of many-sorted logic are applied to axiom sytstems which are naturally many-sorted, these proof procedures appear to be more efficient than the proofs obtained by translating many-sorted into ordinary first order logic. This has been exploited by those interested in symbolic manipulation.

Algebraic Data Types. The purely equational part of many-sorted logic (the theory of many-sorted identities) is the basis of the theory of algebraic data types. Data structures such as stacks, with their push and pop operations, are naturally axiomatizable in the many-sorted logic of identities, and specification and correctness questions can be treated by universal algebra methods stemming from the completeness theorem of Birkhoff for identities (Ehrig and Mahr [1985]).

Uncertain Reasoning. In knowledge based inference systems such as databases and expert systems, the information and the conclusions drawn are not certain. Classical logic deals only with certainty. Statistical inference deals usually with one-step inference based on samples, and often avoids many step inference due to unavailability of large samples. The rules extracted from experts for expert systems are usually not based on statistics and are usually not certain. There is a literature devoted to attaching measures of belief to premises and rules of derivation so as to attach measures of belief to conclusions. Such measures are greatly desired by users of databases and expert systems. There is as yet no convincing way of extracting measures of belief from experts, or applying any calculus for combining such measures. This is an interesting and open area of research. See Duda et al. [1976] (also in Webber and Nilsson [1981]) for a system of Bayesian probabilities widely used for expert systems based on semantic nets.

Non-Monotonic Reasoning. For classical logic the axioms obeyed by the consequence relation "statement ϕ is a consequence of set T of statements" were first described by Tarski in 1930 (see papers III, V, XII in Tarski [1956]). His axioms imply that if $S \subset T$ and ϕ is a consequence of S, then ϕ is a consequence of T. This is the monotonicity property. In classical logic a consequence of a set T of axioms is never withdrawn later due to new axioms being added to T. In contrast, when the entries of a database are regarded as axioms and a consequence is drawn from these axioms, that consequence may later have to be withdrawn. The consequence may say that there is no entry satisfying a query, and later an entry satisfying the query may enter the database when the database is updated. The consequence may say there is an entry satisfying a query, and those entries satisfying the query may have to be withdrawn from the database when the database is updated. The set of answers satisfying a logical query about entries in a database may wax and wane. The rules of inference of classical logic permit a conclusion to be inferred from a set of axioms T only due to certain statements belonging to T, never because certain statements cannot be derived from T. When rules of proof are introduced permitting the latter, then we are dealing with a new definition of proof in a non-monotonic logic. There are different proposals for non-monotonic logics formalizing different intuitive ideas. We mention the logic of default reasoning of Reiter [1980], the use of modal logics by McDermott and Doyle [1980], the papers of McDermott [1982] and Davis [1980], the use of intuitionistic logic by Gabbay [1982] (see also Gabbay in Apt [1985]). Clark [1978] looks at a non-standard negation corresponding to absence of a counterexample in the current database (see Shepherdson [1984, 1985] for examples and criticisms). Semantically defined consequence operations become non-monotonic when the class of acceptable models is restricted, for instance if only minimal models are considered (see McCarthy [1980]). Doyle [1979] proposes a practical truth maintenance system for keeping track of the status of assumptions. See also Doyle [1983].

1.5. Program Correctness.

Proving very large programs correct, sequential or concurrent, is an exciting area of research involving logic and algorithms. Such near disasters as the red-alert warnings created by software errors in NORAD's early warning system have made this a field of vital concern. In 1966 Naur [1969] introduced proofs of correctness. Floyd [1969] attached assertions to the edges of a flaw graph (see Gries [1981]) associated with the program, and suggested that a formal calculus of proof techniques could be provided for proving properties of programs. Hoare developed such a logical calculus, now called Hoare logic, for proving the so-called partial correctness of programs. For a sample of correctness work using first order automated reasoning see Chang and Lee [1973], Clark and Sickel [1977], Clarke and Tärnlund [1979], Hansson and Tärnlund [1979], Hogger [1977, 1978], Apt [1981], and Balogh [1981]. See dynamic logic below as well.

Proofs that concurrent programs behave as specified over all time require demonstrating properties such as invariance (a property continues to hold forever), precedence (one property always precedes another), fairness (every process that should be activated eventually gets its chance to act).

Proofs that programs are correct are now carried out by induction on some measure of depth using semantical or syntactical arguments in one formal system or another. Even for fairly short concurrent programs, proofs are extremely tedious.

These proofs often have many cases. It is difficult to find the right predicates which make the induction valid. One tries to carry out an inductive proof over and over with stronger and stronger induction predicates until a proof is obtained that works.

Correctness proofs for concurrent processes have been given in modal, temporal and dynamic logics. For the use of temporal logic to prove concurrent programs correct, see Manna and Pnueli [1982], Clarke et al. in Apt [1985]. For a survey of other methods, see Barringer [1985]. There is a significant literature on verifying computer communication protocols using such tools as reachability arguments, partial verification, and temporal logic. See Hailpern's paper in Apt [1985] for references.

If one were working on a very large project (like the Strategic Defense Initiative), there might be 10,000 other people writing 20,000,000 other lines of code wich will operate 100,000 different physical systems at the same time. As the army of programmers plods on, one is supposed to prove that what was written performs according to specifications. In 20 years of proofs of correctness, no long complicated concurrent programs have been proved correct. Many purported proofs are false, due to inadvertent omissions amongst a myriad of cases. There is always the worry that an unnoticed case, fired by an unexpected combination of inputs, will start World War III.

There is a hope that logics like those sketched above (or even suitably reworked first order logic) will emerge with efficient inference algorithms for interactive automated reasoning about program correctness. Many groups are interested in this topic. This would lighten the burden of proving program correctness for very large programs, by carrying out all cases and recording them so that none will be missed, and possibly by suggesting predicates for inductions based on stored heuristic rules.

Will proving large programs correct ever be automated? Many workers in the field are certain that their special talents and skills will never be replaced by a machine program based on automated reasoning. But who is willing to admit such a possibility on problems that are not easy? Except for Leibniz, the early analysts who discovered many tricks for integration would never have believed that a mechanical device like MACSYMA (Mathlab Group [1977]) could eliminate drudgery. For that matter, before the advent of the Leibniz-Newton calculus, the mathematicians who published long, hard papers evaluating volumes by infinite summation and geometrical transformations would never have believed that all their work would be replaced by a handful of formulas. No one should dismiss so soon in the history of the subject the possibility of automated verification of large programs. Below we discuss briefly two other possible environments for program verification which arise strictly from logic.

1.6. Modal and Dynamic Logic.

These logics extend first order logic and express notions that cannot be fully expressed in first order logic, but which are useful to understanding the meaning of computer programs. Unlike full higher order logic, these logics are still accessible to analysis using the methods of first order model theory and proof theory. We give a very brief description of modal logic and its connection with computing.

Modal Logic. Modal logic has an operation M ('it is possible that') for building formulas in addition to the others of first order logic. Formal modal logic was developed early in the century on purely philosophical grounds by C. I. Lewis (see Lewis and Langford [1932]). Modal logic was given the "possible worlds" semantics by S. Kripke (see Hughes and Cresswell [1968]). Here is a limited "possible worlds" semantics sufficient to describe the example below. A domain (or model) for modal logic will consist of a set D (as in ordinary first order logic), but in addition a set W (of possible worlds) and a binary relation R on W (xRy is read: 'y is accessible from x'). In first order logic we assign to each relation symbol R a single relation $\mathcal{R}$ on the domain D and to each function symbol f a single function $\mathcal{F}$ on D. In modal logic for each possible world w in W we assign an $\mathcal{F}_w$ and $\mathcal{R}_w$, a possibly different interpretation in each possible world. (One can allow different domains for different possible worlds w). Let $\phi(v_1, \ldots, v_n)$ be a formula which involves the possibility operator M. How is it to be interpreted? The notion that has to be defined is: when does a sequence $(a_1, \ldots, a_n)$ satisfy the logical formula $\sigma(v_1, \ldots, v_n)$ in a possible world w? The crucial case is when ϕ is of the form $(M\sigma)$. Then $(a_1, \ldots, a_n)$ is said to satisfy $(M\sigma)$ in possible world w if and only if for some possible world w' accessible from w, $(a_1, \ldots, a_n)$ satisfies σ.

The simplest connection of modal logic with computer programs is this. Let C be a single command in a computer language such as PASCAL, executed on a deterministic machine. The current state of the machine is determined by the assignments of values in storage locations to variables. Call such an assignment a store. Then the effect of executing command C is to change store w into store w'. Let D be the set of storage locations, let the set W of possible worlds be the set of all stores, let the mapping w to w' be the accessibility relation that command C leads from w to w'. Each command C yields an accessibility relation and a modal operator, and there are a lot of commands. So a simple modal logic with one modal operator is not of much use, and modal logic must be extended for this purpose to have many modal operators. This is what Pratt's version of temporal logic [1976] does. (For early temporal logic see Kamp [1968], Rescher and Urquart [1971].)

Hoare's logic [1969] was designed for specification and proving partial correctness of computer programs. A basic notion in Hoare logic is the construct $A\{G\}B$, meaning that if A holds before the execution of G, then B holds afterwards. (See Gries [1981].) Pratt [1976] was motivated by Hoare [1967] and the Hughes and Cresswell [1972] exposition of modal logic to develop a modal logic of programs in which each command C in a computer language represents a modal necessity operator $[C]$, such that the modal statement $[C]\phi$ expresses that after command C is executed, ϕ holds.

Dynamic logic had as predecessors Engeler's work on algorithmic properties of structures [1967], the algorithmic logic of Salwicki [1970] (see also Banachowski et al. [1977]), and the monadic programming of Constable [1974]. The basic idea of dynamic logic is as follows. At a given stage of execution each variable in a computer program has a value in its domain of possible values. The set of all such assignments is the set of states of the machine. A formula with such variables as free variables will then be true at some stages in the execution, false at others, depending on the current state of the machine. A program beginning in any initial state s, always terminating in a corresponding final state s', can be regarded as a mapping of states to states, and a non-deterministic program of this kind can be regarded as a binary relation among states. Call these input-output programs, and identify each program with the corresponding input-output relation among states. Dynamic logic deals with statements $[p]\phi$ which mean that after program p terminates, ϕ holds. Dynamic logic raises input-output programs to the same status as the predicates of first order logic. In dynamic logic there are operations for constructing new input-output programs from old, along with the usual first order logical operations for building complex formulas out of atomic formulas. The subject has a well-behaved model theory which mixes Kripke modal semantics with the automata theory of regular events. It is sufficiently expressive for specification and correctness problems involving input-output programs. It was not designed for examining properties of their intermediate executions states, or for expressing correctness of non-terminating programs such as operating systems. Other logics, such as process logic, or communicating sequential processes, or temporal logic, or perpetual processes for Horn clause logic deal with these problems. Pnueli [1977] suggested that temporal logic could be used to reason about concurrent programs and operating systems. There are temporal operators such as "p is true sometime in the future", "p is true at the next instant", "p is true at all future times", "q is true at some future time and until then p remains true". Temporal logic deals successfully with proving properties of finite state concurrent programs such as those solving the "dining philosophers" problem. Process logic combines features of both dynamic and temporal logic so as to be able to reason about paths, that is about sequences of intermediate states which represent the course of the computation. For process logic see Kozen and Parikh [1982], for communicating sequential processes based on Scott's continuous operator semantics see Hoare [1985], for perpetual processes based on infinite term algebras for Horn clause logic see Lloyd [1984], for temporal logic see Manna and Pnueli [1981, 1982], Mann and Wolper [1984], Gabbay et al. [1983]. For general dynamic logic, see Pratt [1976, 1980, 1984], Harel [1979, 1984], Harel et al. [1977, 1982, 1983]. For treating processes using temporal logic, see Nguyen et al. in Parikh [1985].

1.7. Automation of Mathematics.

Boyer-Moore Theorem Prover. Boyer-Moore [1979] uses automated classical logic augmented by heuristic proof-finding principles to study mathematician-directed automated proofs in number theory and program verification. Much of their work is devoted to algorithms helpful in completing proofs by induction. When an attempt to complete a proof by induction based on a simple induction hypothesis fails, their algorithms yield more complicated induction hypotheses for a more refined try at completing a proof by induction. Finding such refinements is a principal problem in proving programs correct, which is one reason why the Boyer-Moore environment is promising for computer science.

Constable's PRL. Intuitionistic mathematics was Brouwer's response to the foundations of mathematics paradoxes of Russell and Burali-Forti at the turn of the century. These paradoxes also gave rise to the Hilbert program, see below. Brouwer's proofs were constructive in the sense that if one proves existence of an object, there is inherent in the proof a construction of the object. This is reflected in Kleene's recursive realizability theory for intuitionistic logic. A formal first order logic of intuitionism was developed by Heyting (see Heyting [1966]). There has been extensive development of higher order intuitionistic logic and mathematics, including developments by Bishop, Kleene, Myhill, Martin-Löf, Sanin, Scott, and Troelstra (see Heyting [1966], Bishop [1970], Scott [1970], Troelstra [1973], Brouwer [1975] for references). Since intuitionism uses constuctive methods, it is natural to automate intuitionistic logic and mathematics. Constable's PRL program (Bates and Constable [1982]) is an implementation of the constructive proof methods of the Martin-Löf ([1975], [1982]) higher order intuitionistic logic. Automated intuitionistic mathematics should be able to convert proofs to programs. This gave rise to Programming Logic (Constable and O'Donnell [1978]), and Constable and Bates' PRL program of the early 1980's, and the current Nuprl program (PRL Staff [1985]). This is a mathematician-directed implementation of Martin-Löf's intuitionistic logic for the development of algebra, number theory, and analysis. Constable [1985], building on his experience in writing Nuprl, interprets proofs as expressions which denote evidence, and statements as constructively true when the evidence for them can be computed from their proofs. For other work on the use of the Martin-Löf system for programming see Martin-Löf [1982], Smith [1982], Nordström [1981], Norström and Smith [1983].

Higher Order Logic. Andrews et al. [1982] automated higher order classical logic with some success. Their theorem prover, without real hints, proved Cantor's theorem that $2^X > X$.

There is a hope that logic systems like these will eventually provide "software engineering assistants" for writing verified software.

PART 2. APPLICATIONS ASSOCIATED WITH INCOMPLETENESS.

2.1. Hilbert's Program.

Purely philosophical or foundational questions really do have practical consequences. We take Hilbert's program as an example. The geometric axioms of Euclidean geometry were not completely written out until Hilbert's essay [1902] on the Foundations of Geometry. In his essay the logic was unformalized. He simply took for granted the logic of everyday mathematics. He used multiple interpretations of formal geometry axioms to give proofs of independence and relative consistency. This idea stemmed from earlier nineteenth century proofs that non-Euclidean geometries were consistent if the ordinary geometry of space is consistent, by showing that the non-Euclidean geometries can be interpreted in ordinary geometry, a subject in whose consistency we have confidence.

Hilbert's program of the 1920's was a program to restore confidence in Cantorian mathematics after the paradoxes of Russell and Burali-Forti. Hilbert hoped to justify the consistency of strong systems encompassing Cantorian mathematics, by reducing their consistency to the consistency of very simple systems of arithmetic in which we all have confidence. Hilbert's technical device was to express the proof predicate 'x is the proof of y' of the strong system in the weak system and to show that no contradiction exists in the strong system by proofs carried out in the weak system. There have been arguments for sixty years as to what weak systems he would have permitted. These are arguments about what it means for a proof to be finitary.

What could be more purely philosophical than this program of Hilbert? Yet this is what led to the development of Church's lambda calculus (now widely used in computer science), the first general purpose interpreter for a computer language (the universal Turing machine), the Post production systems (used to write expert systems), and subrecursive hierarchies (pointing the way to computational complexity theory).

2.2. Gödel's Incompleteness Theorem.

What Gödel [1931] did was to show that theories such as set theory or number theory cannot prove their own consistency. But what theories did Gödel's argument apply to? Just those theories whose proof procedures compute the functions required for Gödel's proof of incompleteness. Gödel in a footnote says that perhaps the notion of computable function is an absolute one, that making a formal system richer beyond a certain point may not increase the functions the system can compute by means of its proof procedures regarded as computation procedures. Church proposed his lambda calculus as sufficient to compute all possible intuitively computable functions. The lambda calculus is intended to capture the notion of function as rule. It was a foundation of mathematics alternative to the one provided by set theory. The research workers in the period 1931–1940 were very concerned to show that the notion of recursive function was independent of the formulation of algorithm used. To investigate alternative definitions, Kleene introduced the equation calculus to describe computable functions, as well as the definition of recursive function by primitive recursion and the least number operator, and proved these equivalent to the definition using lambda calculus. Independently, Turing came up with a definition of computation via Turing machines, and Post came up with a definition via Post production systems. All these formalizations of the notion of algorithm were shown to give rise to the same notion of computable function. This has led to nearly universal acceptance of a credo known as Church's thesis, namely that the class of intuitively computable functions is captured by (any one of) these formalisms. Church's lambda calculus, Post's production systems, and Turing machines have all had a profound influence on computer science.

2.3. Church's Lambda Calculus.

Lambda calculus stems from the combinatory logic of Shönfinkel [1924] and Curry [1929, 1930]. Before the 1960's lambda calculus was regarded primarily as an alternative foundation of mathematics. To most logicians it was a side issue. Since that time lambda calculus has been implemented in computer languages and has been used to interpret computer languages. The lambda calculus is based on two primitive notions. One is application of a "function" t to an argument a to get value $t(a)$. The other is abstraction, which forms from a term $t(v)$ the "function" $\big(\lambda x(t(x))\big)$ that assigns to argument a the value $t(a)$. Church intended the lambda calculus as a calculus for constructing and evaluating "functions" thought of as rules which may apply to anything, as distinguished from functions as we currently define them as sets of ordered pairs. In Church's theory, self-application of a "function" t, $t(t)$, can make perfect sense.

Lambda calculus is still not part of the standard logic courses that logicians take, and is certainly not part of the common knowledge of mathematicians. So we describe its notations and axioms briefly (see Church [1932/3, 1941], Church-Rosser [1936], Barendregt [1984] or Stoy [1977] for details). The primitive vocabulary is a list of variables, a symbol λ (the abstractor), and parentheses. Strings are finite sequences from this vocabulary. The set of λ-terms Λ is the smallest set of strings containing the variables such that: (i) whenever Λ contains a string M and x is a variable, Λ also contains $(\lambda x M)$; (ii) whenever Λ contains strings M, N, then Λ also contains the string (MN). Omit outside parentheses, associate λ to the left, and write $\lambda x y.M$ for $\lambda x\big(\lambda y(M)\big)$. The Church-Rosser theory (Church-Rosser [1936], Church [1941]) studies simplifying λ-terms using rewrite rules. Each rewrite rule consists of a pair of expressions $s \to t$ where s, t are λ-terms. Here are the rules.

(1) $\lambda x.M \to \lambda y.M'$, provided y is not free in M and M' is the result of legally

substituting y for x throughout M. (We omit the definition of legal.)

(2) $(\lambda x.M)N \to M'$, where M' is the result of legally substituting N for x throughout M.

(3) $\lambda x.Mx \to M$, provided x does not occur free in M.

Each of these rules has a converse with left and right hand sides reversed. The general rewrite rule of inference says that given a rewrite rule $s \to t$ and a term ϕ and an occurrence of s in ϕ, then ϕ can be rewritten ϕ', where ϕ' is the result of substituting t for s at its occurrence in ϕ. Also the s to which a rewrite rule applies is called a redex.

We say s and t are convertible if rules (1), (2), (3) and their converses allow us to rewrite s as t. We say s is reducible to t if we can rewrite s as t using only (1), (2), (3), but not their converses. An expression with no redex within it (no possible simplification) is called a normal form expression. Not every expression has a normal form, for example $(\lambda x.xx)(\lambda x.xx)$.

Church-Rosser Theorem. *If A can be converted to B, then there is a term C such that A and B can both be reduced to C.*

So if there is a normal form, it is independent of the reductions used. The standard textbooks are Hindley et al. [1972], Stenlund [1972], and Barendregt [1984].

Knuth-Bendix Rewrite Rules. Knuth-Bendix [1970] combined the Church-Rosser reduction idea with unification to give a theory of rewrite rules for expressions in general, not just lambda terms. A rewrite system is a finite set of ordered pairs (s,t) of expressions, representing rewrite rules $s \to t$. The basic rewriting rule of inference again allows substitution of t for s in ϕ when s occurs as a subexpression of ϕ. Application of this rule may be limited to the use of most general unifiers if such a notion is available in the context. There is an equivalence relation like conversion generated by rewrite rules and their converses, and a notion of reduction using only the rewrite rules but not their converses. There are many situations in mathematics and computer science when symbolic expressions reduce by rewrite rules to a unique canonical form which is better for computation. Often there is a non-negative integer value associated with an application of rewrite rules applied to ϕ to obtain ϕ', with the integer value of the rewrite ϕ' less than the integer value of the original ϕ. Frequently expressions with minimum integer value are in canonical form. The confluence property of a set of rewrite rules is that whenever an expression can be rewritten in two different ways, these have a common rewriting. Knuth-Bendix [1970] observed that naturally occurring rewrite rules can often be refined to a set of confluent rules. They gave an algorithm which, operating on a set of rewrite rules, often replaces a set of rewrite rules by a confluent set. Confluent rules have the Church-Rosser property that any two convertible expressions have a common rewrite, and in the presence of an integer valuation such as mentioned above this gives canonical forms. See Huet and Oppen [1980] for references, see Plotkin [1972] for an important contribution, see Jouannaud [1985] for a collection of recent rewrite rule papers.

2.4. Lambda Calculus and Programming Languages.

When "functions" are executed by programs on computers, the machines do not recognize type distinctions unless this is deliberately enforced. "Functions" can apply to themselves. This is a natural area of application for the lambda calculus both as a calculus for interpreting computer programs and as a calculus which can be implemented by computer languages based on evaluation of functions.

LISP. One of the oldest computer languages, McCarthy's LISP, (McCarthy [1960], [1962]) was designed to process lists of symbols, and therefore to support symbolic manipulation. Most of the famous AI programs such as MACSYMA were originally composed in LISP. Hidden under the hood of the surface notation for LISP is the engine of Church's lambda terms. LISP incorporates the ability to handle functions as objects and to pass functions as parameters. For mathematicians who know nothing about LISP, an exceedingly clear introduction to LISP programming is Wilensky [1984]. If one wants to know how LISP implements lambda expressions, this is explained well in Henderson [1980]. GEDANKEN is a language explicitly modeled on the lambda calculus (Reynolds [1970]). For the use of the lambda calculus to analyze programming languages, see Landin [1964, 1965, 1966a, 1966b], Morris [1968], Gordon [1973], Burge [1978], and Gordon [1979], Gordon et al. [1979].

2.5. Continuous Operators and Programming Languages.

Enumeration Operators. In recursion theory not only is there a theory of recursive functions on integers to integers, but there are theories of recursive operators on sets of integers to sets of integers. One of these is the theory of enumeration operators (Kleene [1950], Kreisel [1959], Nerode [1959]). Let $P(\omega)$ be the set of all subsets of the integers ω. The enumeration operators on $P(\omega)$ to $P(\omega)$ are the continuous operators relative to the topology in which an open set is the union of sets of the form U_F, where U_F consists of all supersets of a fixed finite subset F of ω. These are the operators ϕ for which the question whether x is in $\phi(A)$ is determined by whether one of a countable number of finite sets F is contained in A. In 1969 Scott found that simple models of the lambda calculus could be made up using continuous operators. The problem he solved was making sense of the self-reference that forming $t(t)$ implies about "functions" t in the lambda calculus, and he solved this difficulty by restricting attention to a class of continuous operators (the enumeration operators) that had arisen in recursion theory.

The Powerdomain $P(\omega)$. We cannot reproduce the whole idea, but here is a form of Scott's observation. Let j be the pairing functions matching $\omega \times \omega$ with ω given by $j(x,y) = \big((1/2)(n+m)(n+m+1)\big) + m$, let $\{F_n\}$ be the list of all finite subsets of ω in which F_n is the set of exponents in the binary expansion of n. Each continuous operator $\phi : P(\omega) \to P(\omega)$ is determined by the set

$$\text{graph}(\phi) = [j(n,m) \text{ such that } m \text{ is in } \phi(F_n)].$$

Conversely, every set of integers A determines a continuous operator $\text{fun}(A) : P(\omega) \to P(\omega)$ defined for U in $P(\omega)$ by

$$\text{fun}(U)(X) = \{m \mid \text{for some } F_n \subset X, \ j(n,m) \in U\}.$$

Define application $(XY) = \text{fun}(X)(Y)$. For continuous $f(x,\vec{Y}) : P(\omega) \times P(\omega)^k \to P(\omega)$ define $\big(\lambda x f(x,\vec{Y})\big) = G(\lambda x f)$. This can be used to define interpretations of λ-terms relative to a valuation of variables in $P(\omega)$. Scott was able to define models of the lambda calculus as subsystems of $P(\omega)$ (see Scott [1970], [1976], Plotkin [1976], or Stoy [1977] for details).

This was followed by the development of many models for typed and untyped lambda calculus. This not only cleaned up the semantics of the lambda calculus, but meant also that the earlier 1960's work of interpreting programs as computations in the lambda calculus could now be replaced by a mathematical semantics using continuous operators. This idea of Scott is the basis of the Scott-Strachey theory of denotational semantics of programming languages (Scott-Strachey [1971] and Stoy [1981]). For a few later applications of related ideas see Milner [1980] and Hoare [1985, §§ 2.8, 3.9]. Other frameworks used now are continuous lattices and complete partial orderings (cpo's). The latter are partial orderings with a smallest element, in which directed subsets have least upper bounds. Another decisive idea is the interpretation of cartesian closed categories as models of the lambda calculus. Topos theory in its guise as the algebra of intuitionistic and constructive mathematics has an important role to play in the semantics of computer languages and in suggesting new high level languages. See Barendregt [1984] for the algebra, topology, and model theory of lambda calculus.

2.6. Universal Turing Machines.

Turing's paper [1936] introduces Turing machines. Each such machine $\mathcal{M}$ operates on an infinite sequence of adjacent initially blank squares. Each square is capable of being printed with one letter from a fixed finite alphabet, including a blank symbol. The machine has a read-write head capable of being in one of a finite number of internal states. It can read or write only on one square at a time. The program of Turing machine $\mathcal{M}$ consists of three tables, describing three things that $\mathcal{M}$ does at each move. At a given move, on the basis of the current state of the read-write head and the current symbol being read on the scanned tape square, M is instructed by the tables to move right or left one square, to overprint the currently scanned symbol, and to go into a new internal state. Although the original description of these machines looks limited, nonetheless it was the basis of the first convincing philosophical argument that any function of integers that is calculable in the intuitive sense can be calculated by an algorithm, a Turing program. One impact of Turing machines is that the current theory of computational complexity is based on resource limited Turing machines—for example, polynomial time bounded deterministic or non-deterministic machines defining P and NP respectively (see Garey and Johnson [1979]).

But Turing's paper [1936] did not consist merely of a definition of Turing machines and a philosophical argument. It contained the construction of a universal Turing machine U. This construction is the basis of recursive function theory. The whole construction, regarded as very specialized thirty years ago, is now part of the standard undergraduate junior level training of every computer science major (see Hopcroft and Ullman [1979]). But there is another importance to this construction. The universal Turing machine was also the first general purpose interpreter. What an interpreter (running on a sequential computer U) does is to take a program M which has been read into the memory of U, and (at each stage of the computation), fetch one instruction from the program M and have U execute it. U simulates what M says to do, one instruction at a time. That is exactly what the universal Turing machine U does with arbitrary Turing programs M. A Turing machine program M (the three tables above) is written on the tape (memory) of the universal machine U. At each stage of the computation the read-write head of U fetches an instruction from this program of M on tape (in memory) and U executes that instruction in simulated form. U is the first general purpose interpreter, in the sense that it intentionally incorporates enough interpretive power to interpret programs that represent all possible computable functions (as we know them from Church's thesis). Every interpreter for every major higher level language nowadays is (with very slight modifications) capable of the same universality, if given potentially infinite storage corresponding to Turing's tape (as much RAM and hard disk space as needed at any time). Von Neumann is usually given credit for the basic design of sequential machines from control, memory, and arithmetic units. The arithmetic units were absent from Turing's formulation, since he was not interested in efficiency in that paper and that is what arithmetic units add. Some have questioned how much Von Neumann knew about Turing's paper, or whether he used Turing's ideas. Based on Von Neumann's deep knowledge of Hilbert's program, on his early significant contributions to logic, and on his friendship with Gödel (who regarded Turing's work as very significant), on his later work on tesselation models and self-reproducing machines (which are extensions of the universal Turing machine construction), it is hard to doubt that this was a source for his ideas.

Self-Reproducing Machines. We cannot resist an aside on Von Neumann's model for self-reproducing machines (Von Neumann [1966], Burks [1970]). In the early 1960's, many people (Myhill and Thatcher, for example) worked out the consequences of adding to the self-reproducing machine the blueprint of an extra desired task, which would then be carried out as a side product of self-reproduction. This would allow harvesting a byproduct if energy is provided for the process. I was on a panel twenty years ago with a famous biochemist. When I mentioned the Von Neumann idea as a possible prototype model for reproduction which might shed light on biological mechanisms, the biochemist was certain that in genetic reproduction the complex biochemical reactions could not be conceptualized digitally. Now, twenty years later, the gene splicers routinely add genes to organisms and routinely produce useful byproducts. The current partial models for this are practically as digital as Von Neumann's.

His cellular automata may yet have other uses. They appear in Conway's game of life (see Burlekamp et al. [1982]) and have been used by Wolfram as models for the evolution of physical processes. In these cellular automata models the computation of new state from old is boolean, no real numbers or floating point operations are involved. These computations are naturally parallel. They represent an alternative to the traditional physical models which use difference equations and floating point arithmetic.

2.7. Post Production Systems.

E. L. Post worked quite hard in the 1920's and 30's on a method of defining formal systems. Now the production systems he invented are used widely in computer science and linguistics. Post developed the notion of a production system as a formalization of the most general concept of a formal system. His idea was that, with trivial rewriting, all formal systems are production systems.

Let A^* be all strings (finite sequences) from A, with elements of A considered as an alphabet. Let $V = \{\alpha_1, \alpha_2, \ldots\}$ be variables ranging over A^*. We call elements of V string variables. Then a metastring is a string from $(A \cup V)^*$. A production P is of the form

$$m_1, \ldots, m_k \to m_{k+1},$$

where $m_1, \ldots, m_{k+1}$ are metastrings. An instance of a production is obtained by substituting strings for its metastring variables. A Post production system P consists of a finite set of productions and a finite set of strings (called its axioms). The set T of theorems of P is the smallest set of strings from A^* containing the axioms and such that whenever string $m'_1, \ldots, m'_k$ are in T and there is a production which has as an instance (by substitution of strings for string variables)

$$m'_1, \ldots, m'_k \rightarrow m'_{k+1},$$

then m'_{k+1} is in T. This generalizes every definition by recursion of any set of expressions in any language ever given. For Post, production systems were purely formal. They have, however, many interesting interpretations, each with a different sphere of applications. Production systems are the basis for defining classes of mathematical grammars studied by mathematical linguists.

OPS5. Post's systems are also the basis for production system programming languages (such as Carnegie-Mellon's OPS5) used to study cognitive science models and to write expert systems. In OPS5 a wrinkle from cognitive systems modelling is added to Post production systems. This is the idea of modelling short term and long term memory, and how they may communicate. Axiom strings and production rules are stored separately, strings in a short term "working memory" area, and production rules in a long term "production memory" area. A production rule fires if the left hand side instantiates to a string in working memory. However, the instantiated right hand side of the production rule is interpreted here not merely as a string as in Post, but as a set of instructions for actions to be performed modifying strings in working memory. This is how long and short term memory communicate in OPS5. The interpretation of the right hand side of fired production rules as prescriptions for subtracting from or otherwise modifying the set of temporary "theorems" of working memory is a new feature not present in Post. One never loses theorems in Post production systems, the way one does with short term memory here. The OPS5 implementation of short term memory as working memory for strings is a hybrid of Post's production systems and Turing machines. Turing machines store strings on tape to be both read and modified by the program in the Turing machine readwrite head. The Turing program corresponds roughly to the productions in the Post production systems and to production memory in OPS5.

The proofs that the various definitions of computable functions were equivalent were themselves algorithms for translating programs from one language to another, from lambda calculus to Post production systems, from Post production systems to Turing machines, from Turing machines to Kleene recursion equations, etc. Every one of these formalisms has been made the basis of a high level computer language. For example, see O'Donnell [1985] for recursion equations as a programming language. The translation algorithms can be interpreted as early cross-interpreters, and in a couple of cases as compilers.

2.8. Search Trees and Pattern Matching.

Some common features of all systems of automation of reasoning can be laid out. One must specify a labelled tree of possible derivations, how this tree is to be searched, and how patterns are to be matched during this search. One wants efficient pattern matching algorithms and efficient tree searching algorithms. The simplest search procedures are forward and backward chaining. In logic, forward chaining corresponds to the way proofs are generated in what logicians call Hilbert-type systems by applying the rules of inference in all possible ways to the axioms, and then repeating this application to what has been obtained, etc. This is what is employed in OPS5. This is very convenient for expert systems based on cross-classification such as fault diagnosis or advisors for filling out income tax forms. Backward chaining corresponds to the way proofs are written in Gentzen systems of natural deduction, starting with the desired conclusion and seeing how the conclusion could have arisen as the last line of a proof from previous lines,then looking at the possible previous lines, and repeating till all is justified with axioms. If one knows the conclusion one is trying to reach, then searching for a justification of it from the premises is an appropriate application of backward chaining search. PROLOG uses backward chaining. Forward chaining is often characterized as data driven, backward chaining as goal driven. Obviously any system of logic can be implemented either way, depending on what use one wants to make of it.

Now for pattern matching. One of the reasons for the practicality of a production system like OPS5 is the simplicity of pattern matching required to check whether or not a string is an instance of a metastring—i.e., whether a production rule fires based on a string in working memory. This pattern matching is all that is needed in many commonly encountered expert systems, and is less resource intensive than a unifier of terms for PROLOG. On the other hand, for many comon PROLOG applications only a trivial level of unification is required anyway. It is only the combination of logic (for knowledge representation and inference) and algorithms (for search and pattern matching) that makes useful automated reasoning systems. Developing logic systems for knowledge representation and developing control structures for logical searches are subjects in their infancy. See Nilsson [1980].

FINAL REMARKS.

There is now a *Journal of Automated Reasoning* (edited by J. A. Robinson) and a *Journal of Logic Programming*. There is an annual international conference called LICS (Logic in Computer Science). The name of one of the principal logic journals, the *Annals of Mathematical Logic*, was recently changed to the *Annals of Pure and Applied Logic* and has computer scientists on its editorial board. The *Journal of Symbolic Logic*, founded jointly by mathematicians and philosophers, now has computer scientists on its editorial board and encourages papers in logic related to computer science. Many mathematical logic Ph.D.'s are finding employment in computer science departments, or are doing applied logic (such as program verification) in government and industrial research installations.

Most existing applications of logic have used ideas well developed by the 1950's. Indeed, that was the principle by which this lecture was organized. Since then, many deep results have been developed in set theory, recursion theory, proof theory, and model theory. Will ideas from these later developments play a significant role in future applications of mathematical logic? We think so.

Acknowledgment. This research was supported in part by NSF grant MCS-8301850.

References.

A. V. Aho and J. D. Ullman [1979], "Universality of data retrieval languages," *Proc. Sixth ACM Symp. on POPL*, 110–117.

J. F. Allen [1981], "An interval-based representation of temporal knowledge," *Proc. Seventh IJCAI*, Vancouver, 221–226.

J. F. Allen and J. A. Koomen [1983], "Planning using a temporal world model," *Proc. Eighth IJCAI*, 741–747.

H. Andreka and I. Nemeti [1976], "The generalized completeness of Horn predicate logic as a programming language," *DAI Report 21*, University of Edinburgh.

P. B. Andrews [1982], "A look at TPS," *Sixth Conf. on Automated Deduction*, Lecture Notes in Computer Science 138, Springer-Verlag, New York.

K. Apt [1981], "Ten years of Hoare logic, a survey (part 1)," *ACM Trans. on Prog. Languages and Systems*, **3**, 431–483.

K. Apt [1985], *Logics and Models of Concurrent Systems* (K. Apt, ed.), NATO ASI Series, Springer-Verlag, Berlin.

K. Apt and M. H. van Emden [1982], "Contributions to the theory of logic programming," *J. ACM*, to appear.

K. Balogh [1975], "Software applications of mathematical logic" (Hungarian), *Proc. Conf. Prog. Systems*, Szeged, Hungary, 26–44.

K. Balogh [1981], "On an interactive program verifier for PROLOG programs," *Proc. of Coll. on Math. Logic in Programming*, Salgotarjan, Hungary, 1978; republished (B. Domoki and T. Gergely, eds.), North Holland Pub. Co., Amsterdam.

L. Banachowski, A. Kreczmar, G. Mirkowska, H. Rasiowa and A. Salwicki [1977], "An introduction to algorithmic logic," in *Mathematical Foundations of the Theory of Programs* (Mazurkiewitz and Pawlak, eds.), Warsaw, 7–99.

H. P. Barendregt [1984], *The Lambda Calculus* (revised edition), Studies in Logic 103, North Holland, Amsterdam.

H. Barringer [1985], *A Survey of Verification Techniques for Parallel Programs*, Lecture Notes in Computer Science 191, Springer-Verlag, Berlin.

J. L. Bates and R. L. Constable [1982], "Definition of micro-PRL," *Technical Report TR 82-492*, Dept. Computer Science, Cornell University, Ithaca, New York.

J. L. Bates and R. Constable [1985], "Proofs as programs," *ACM Trans. on Programming Languages and Systems*, **7**.

M. Beeson [1984], "Proving programs and programming proofs," *Proc. Seventh Intl. Cong. Logic, Phil., and Method. of Sci.*, North Holland, Amsterdam.

E. Berlekamp, J. Conway and R. Guy [1982], Chapter 25 of *Winning Ways*, Academic Press, New York.

J. Van Benthem [1983], *The Logic of Time*, Reidel.

W. W. Bledsoe [1977], "Non-resolution theorem proving" *Art. Intell*, **9**, 1–35.

W. Bledsoe and P. W. Loveland [1985], *Special Session on Automatic Theorem Proving*, Cont. Math. 29, Amer. Math. Soc., Providence, Rhode Island.

M. Bergmann and H. Kanoui [1973], "Application of mechanical theorem proving to symbolic calculus," *Third Intl. Symp. of Advanced Computing Methods in Theoretical Physics*, C.N.R.S., Marseille.

E. Bishop [1967], *Foundations of Constructive Analysis*, McGraw Hill, New York.

E. Bishop [1970], "Mathematics as a numerical language," in *Intuitionism and Proof Theory* (J. Myhill et al., eds.), North Holland, Amsterdam.

D. G. Bobrow [1980], Special Issue on Non-Monotonic Logic (K. Bobrow, ed.), *Art. Intell.*, **13**.

G. Boole [1854], *Laws of Thought*; in *Collected Logical Works*, Open Court, La Salle, Illinois, 1952.

K. A. Bowen and R. A. Kowalski [1982], "Amalgamating language and metalanguage in logic programming," in Clark and Tärnlund.

R. S. Boyer and J. S. Moore [1973], "Proving theorems about LISP functions," *Proc. IJCAI3* (N. Nilsson, ed.).

R. S. Boyer and J. S. Moore [1979], *A Computational Logic*, Academic Press, New York.

R. J. Brachman and H. J. Levesque [1985], *Readings in Knowledge Representation*, Morgan Kauffman, Los Altos, California.

B. Bradley and M. Schwartz [1979], *Possible Worlds*, Oxford.

L. E. J. Brouwer [1975], *Collected Works 1* (A. Heyting, ed.), North Holland, Amsterdam.

Lee Brownston, R. Farrell, E. Kant and N. Martin [1985], *Programming Expert Systems in OPS5*, Addison-Wesley, Reading, Massachusetts.

de Bruijn [1980], "A survey of Project Automath," in *To H. B. Curry: Essays on Combinatory Logic, Calculus, Lambda Calculus, and Proof Theory* (J. P. Seldin and J. R. Hindley, eds.), Academic Press, London.

M. E. Bruynooghe [1976], "An interpreter for predicate logic programs I," *Report CW10*, Appl. Math. and Prog. Division, Katholieke Universiteit, Leuven, Belgium.

A. Bundy [1983], *The Computer Modelling of Mathematical Reasoning*, Academic Press.

W. Burge [1978], *Recursive Programming Techniques*, Addison-Wesley, Reading, Massachusetts.

A. W. Burks [1970], *Essays on Cellular Automata*, University of Illinois Press.

R. M. Burstall and J. Darlington [1977], "A transformation system for developing recursive programs," *J. ACM*, **24**, 44–67.

J. A. Campbell [1984], *Implementations of PROLOG*, Wiley, New York.

C. L. Chang and R. Lee [1973], *Symbolic Logic and Mechanical Theorem Proving*, Academic Press, New York.

A. Church [1932/33], "A set of postulates for the foundations of logic," *Ann. Math.*, **33**, 346–366; second paper *ibid.*, **34**(1933), 839–864.

A. Church [1941], *The Calculi of Lambda-Conversion*, Ann. Math. Studies 6, Princeton University Press, Princeton, New Jersey.

A. Church and J. B. Rosser [1936], "Some properties of conversion," *Trans. Amer. Math. Soc.*, **39**, 472–482.

K. L. Clark [1978], "Negation as failure," in Gallaire and Minker.

K. L. Clark and M. H. van Emden [1981], "Consequence verification of flowcharts," *IEEE Trans. on Software Engineering,* **SE-7**, 52–60.

K. L. Clark and F. G. McCabe [1982], "PROLOG: a language for implementing expert systems," in *Machine Intelligence 10* (J. E. Hayes, D. Michie and Y. H. Pao, eds.), Ellis Horwood Ltd., Chichester, England, 455–470.

K. L. Clark and S. Sickel [1977], "Predicate logic: a calculus for deriving programs," *Proc. Fifth IJCAI,* Cambridge, Massachusetts.

K. L. Clark and S.-A. Tärnlund [1977], "A first order theory of data and programs," *Proc. IFIP-77,* Toronto, North Holland, Amsterdam, 939–944.

K. L. Clark and S.-A. Tärnlund [1982], *Logic Programming,* APIC Studies in Data Processing 16, Academic Press, London.

W. F. Clocksin and C. S. Mellish [1984], *Programming in PROLOG,* Springer-Verlag, Berlin.

E. F. Codd [1970], "A relational model of data for large shared data banks," *Comm. ACM,* **13**, 377–387.

E. F. Codd [1971], "A database sublanguage founded on the relational calculus," *Proc. ACM SIGFIDET Workshop on Data Description, Access, and Control* (Codd and Dean, eds.), ACM, San Diego, 35–68.

E. F. Codd [1972], "Relational completeness of data base sublanguages," in *Data Base Systems* (R. Rustin, ed.), Courant Comp. Symp. 6, Prentice Hall, Englewood Cliffs, New Jersey, 65–98.

A. Colmerauer, H. Kanoui, R. Pasero and P. Roussel [1972], "Un système de communication homme–machine en Français," *Preliminary Report,* Art. Intell. Group of Aix-Marseille, Luminy, France.

A. Colmerauer, "Metamorphosis grammars," in *Natural Language Communication with Computers* (L. Bolc, ed.), Lecture Notes in Computer Science 63, Springer-Verlag, Berlin, 133–189.

A. Colmerauer [1979], "Sur les bases théoriques de PROLOG," Group Programmation et Languages AFCET, Division Théorique et Technique de l'Informatique 9.

A. Colmerauer, H. Kanoui and M. van Canegham [1981], "Last steps toward an ultimate PROLOG," *Proc. Seventh IJCAI,* Vancouver.

R. L. Constable [1971], "Constructive mathematics and automatic programming," *Proc. IFIP Congress,* Ljubljana, 229–233.

R. L. Constable [1983], "Partial functions in constructive formal theories," *Proc. Sixth G. I. Conf.,* Lecture Notes in Computer Science 135, Springer-Verlag, Berlin.

R. L. Constable [1985], "Constructive mathematics as a programming logic I: some principles of theory," *Annals of Discrete Mathematics,* **2**, 21–38.

R. L. Constable [1985], "The semantics of evidence," *Technical Report TR 85-684,* Dept. Computer Science, Cornell University, Ithaca, New York.

R. L. Constable [1985], "Investigations of type theory in programming logics and intelligent systems," *Technical Report TR 85-701,* Dept. Computer Science, Cornell University, Ithaca, New York.

R. L. Constable and M. J. O'Donnell [1978], *A Programming Logic,* Winthrop, Cambridge, Massachusetts.

H. B. Curry [1929], "An analysis of logical substitution," *Am. J. Math.,* **51**, 363–384.

H. B. Curry [1930], "Grundlagen der Kombinatorischen Logik," *Am. J. Math,* **52**, 509–536, 789–834.

V. Dahl [1981], "On data base system development through logic," Research Report, Dept. of Mathematics, Faculty of Exact Sciences, University of Buenos Aires, Argentina.

M. Davis [1980], "The mathematics of non-monotonic reasoning," *Art. Intell.*, **13**, 73–80.

M. Davis and H. Putnam [1960], "A computing procedure for quantification theory," *JACM*, **7**, 201–215.

E. W. Dijkstra [1976], *A Discipline of Programming*, Prentice Hall, Englewood Cliffs, New Jersey.

J. Doyle [1979], "A truth maintenance system," *Art. Intell.*, **12**, 231–272.

J. Doyle [1983], "The ins and outs of reason maintenance," *IJCAI*, **8**, 349–351.

R. O. Duda, P. E. Hart and N. J. Nilsson [1976], "Subjective Bayesian methods for rule-based inference systems," *Proc. Nat. Comp. Conf.* (AFIPS Conference Proc. 45), 1075–1082.

H. Ehrig and B. Mahr [1985], *Fundamentals of Algebraic Specification*, Springer-Verlag, Berlin.

E. Engeler [1967], "Algorithmic properties of structures," *Math. Syst. Theory*, **1**, 183–195.

L. Fox [1966], *Advances in Programming and Non-Numerical Computation*, Pergamon Press, London.

M. J. Fischer and R. E. Ladner [1978], "Propositional dynamic logic of regular programs," *J. of Comp. Sci. and Sys. Sci.*, **18**, 194–211.

R. Floyd [1967], "Assigning meanings to programs," in *Mathematical Aspects of Computer Science* (J. T. Schwartz, ed.), Proc. Symp. Appl. Math. XIX, Amer. Math. Soc., Providence, Rhode Island, 19–32.

D. Gabbay [1982], "Intuitionistic basis for non-monotonic logic," *Proc. Conf. on Automated Deduction*, Lecture Notes in Computer Science 6, Springer-Verlag, Berlin.

D. Gabbay and F. Guenthner [1983/4], *Handbook of Philosophical Logic*.

D. Gabbay, A. Pnueli, S. Shelah and J. Stain [1980], "On the temporal logic of programs," *Proc. 18th Symp. on Math. Found. of Computer Science*, Lecture Notes in Computer Science 54, Springer-Verlag, Berlin, 112–152.

H. Gallaire [1981], "The impact of logic on databases," *Proc. Seventh Conf. on Very Large Databases*, Cannes, France.

H. Gallaire and C. Lasserre [1979], "Controlling knowledge deduction in a declarative approach," *Proc. IJCAI 6*, Tokyo.

H. Gallaire and C. Lasserre [1980], "A control metalanguage for logic programming," in Tärnlund [1980].

H. Gallaire and J. Minker [1978], *Logic and Databases*, Plenum Press, New York.

H. Gallaire, J. Minker and J. M. Nicholas [1980], *Advances in Database Theory*, vol. 1, Plenum Press, New York.

M. R. Garey and D. S. Johnson [1979], *Computers and Intractability: A Guide to the Theory of $\mathcal{N}P$-Completeness*, Freeman, New York.

G. E. Gentzen [1969], *The Collected Works of Gerhard Gentzen* (M. E. Szabo, ed.), North Holland, Amsterdam.

K. Gödel [1930], "Die Vollständigkeit der Axiome des Logischen Functionenkalkuls," *Monat. Math. Phys.*, **37**, 349–360, in van Heijenoort.

K. Gödel [1931], "Ueber Formal Unentscheidbare Sätze der Principia Mathematica und Verwandter Systeme, I," *Monatsh. Math.*, **38**, 173–198, in van Heijenoort.

A. Goodal [1985], *The Guide to Expert Systems*, Learned Information (Europe), Besselsleigh Rd. Abingdon, Oxford, England.

M. J. C. Gordon [1973], *Evaluation and Denotation of Pure LISP, A Worked Example in Semantics*, Ph.D. Diss., University of Edinburgh.

M. J. C. Gordon [1979], *The Denotational Description of Programming Languages*, Springer-Verlag, Berlin.

M. J. C. Gordon, R. Milner and C. Wadsworth [1979], *Edinburgh LCF. A Mechanical Logic of Computation*, University of Edinburgh.

C. C. Green [1969a], *The Application of Theorem Proving Question-Ansewring Systems*, Ph.D. Diss., Stanford.

C. C. Green [1969b], "The application of theorem proving to problem solving," *Proc. IJCAI 1*, Washington, D.C., 219–240.

C. C. Green [1969c], "Theorem proving by resolution as a basis for question-answering systems," in *Machine Intelligence 4*, Edinburgh University Press, 183–205.

Peter Grey [1984], *Logic, Algebra, and Databases*, Wiley, New York.

D. Gries [1981], *The Science of Programming*, Springer-Verlag, Berlin.

Y. Gurevich [1985], "Logic and the challenge of computer science," *Technical Report CRL-TR-10-85*, University of Michigan Computing Laboratory.

J. Halpern, Z. Manna and B. Moszkowski [1983], "A hardware semantics based on temporal intervals," *Proc. 19th Intl. Coll. on Automata, Languages, and Programming*, Lecture Notes in Computer Science 54, 278–292.

A. Hansson and S. A. Tärnlund [1979], "A natural programming calculus," *Proc. IJCAI 6*, Tokyo.

D. Harel [1979], *First Order Dynamic Logic*, Lecture Notes in Computer Science 68, Springer-Verlag, Berlin.

D. Harel [1984], "Dynamic logic," in Gabbay and Guenthner.

D. Harel, D. Kozen and R. Parikh [1982], "Process logic: expressiveness, decidability, completeness," *J. Comp. Sys. Sci.*, **25**, 144–170.

D. Harel, A. Meyer and V. R. Pratt [1977], "Computability and completeness in the logic of programs," *Ninth ACM Symp. Theory of Comp.*, 261–268.

D. Harel, A. Pnueli and J. Stavi [1983], "Propositional dynamic logic of nonregular programs," *J. Comp. Sys.*, **16**, 222–243.

P. J. Hayes [1977], "In defense of logic," *IJCAI*, **5**, 559–565.

P. Henderson [1980], *Functional Programming: Application and Implementation*, Prentice Hall, Englewood Cliffs, New Jersey.

J. Herbrand [1930], "Researches in the theory of demonstration, investigations in proof theory," in van Heijenoort.

A. Heyting [1966], *Intuitionism*, North Holland, Amsterdam.

D. Hilbert [1902], *The Foundations of Geometry*, Open Court Pub. Co., La Salle, Illinois.

R. Hill [1974], "Lush-resolution and its completeness," *DCL Memo 78*, Dept. Art. Intell., University of Edinburgh.

J. R. Hindley [1983], "The completeness for typed lambda terms," *Theor. Comp. Sci.*, **22**, 1–17.

J. R. Hindley, B. Lercher and J. P. Seldin [1972], *Introduction to Combinatory Logic*, Cambridge University Press.

C. A. R. Hoare [1969], "An axiomatic basis for computer programming," *Comm. Assoc. Comp. Mach.*, **12**, 576–583.

C. A. R. Hoare [1985], *Communicating Sequential Processes*, Prentice Hall, Englewood Cliffs, New Jersey.

C. J. Hogger [1978], "Program synthesis in predicate logic," *Proc. AISB/GI Conf. on AI*, Hamburg, Germany.

C. J. Hogger [1982], "Logic programming and program verification," in *Programming Technology: State of the Art Report* (P. J. L. Wallis, ed.), Pergamon Infotech Ltd., Maidenhead, England.

C. J. Hogger [1984], *Introduction to Logic Programming*, Academic Press.

J. E. Hopcroft and J. D. Ullman [1979], *Introduction to Automata Theory, Languages, and Computation*, Addison-Wesley, Reading, Massachusetts.

G. Huet [1974], "A unification algorithm for typed lambda calculus," *Note de Travail A055*, Inst. de Rech. 250, Lab. de Rech. en Informatique et d'Automatique.

G. Huet [1977], "Confluent reductions, abstract properties, and applications to term rewriting systems," *18th IEEE Symp. on the Found. of Comp. Sci.*, 30–45.

G. Huet and D. C. Oppen [1980], "Equations and rewrite rules: a survey," in *Formal Languages: Perspectives and Open Problems* (R. Book, ed.), Academic Press.

G. E. Hughes and M. J. Cresswell [1968], *An Introduction to Modal Logic*, Methuen, London.

N. Immerman [1982], "Relational queries computable in polynomial time," *14th ACM Symp. Theor. of Comp.*, 147–152.

B. E. Jacobs [1985], *Applied Database Logic*, vol. 1, Prentice Hall, Englewood Cliffs, New Jersey.

J.-P. Jouannaud [1985], *Rewriting Techniques and Applications* (J.-P. Jouannaud, ed.), Lecture Notes in Computer Science 202, Springer-Verlag, Berlin.

K. Kahn [1981], "Uniform – a language based on unification which unifies (much of) LISP, PROLOG, and ACT1," *IJCAI-81*.

W. Kamp [1968], *Tense Logics and the Theory of Linear Order*, Ph.D. Diss., UCLA.

S. C. Kleene [1936], "λ-definability and recursiveness," *Duke Math. J.*, **2**, 340–353.

S. C. Kleene [1950], *Introduction to Metamathematics*, Van Nostrand, New York.

F. Kluzniak and S. Szpakowicz [1985], *PROLOG for Programmers*, Academic Press.

D. E. Knuth and P. B. Bendix [1970], "Simple word problems in universal algebra," in *Computational Problems in Abstract Algebra* (Leech, ed.), 263–297.

R. A. Kowalski [1970], *Studies in the Completeness and Efficiency of Theorem Proving by Resolution*, Ph.D. Diss., University of Edinburgh.

R. A. Kowalski [1971], "Linear resolution with selector function," *Art. Intell.*, **2**, 227–260.

R. A. Kowalski [1974a], "Logic for problem solving," *DCL Memo 75*, Dept. Art. Intell., University of Edinburgh.

R. A. Kowalski [1974b], "Predicate logic as a programming language," *Proc. IFIP-74*, North Holland, Amsterdam, 569–574.

R. A. Kowalski [1978], "Logic for data descriptions," in Gallaire and Minker.

R. A. Kowalski [1979a], "Algorithm = Logic + Control," *Comm. ACM*, **22**, 424–431.

R. A. Kowalski [1979b], *Logic for Problem Solving*, Art. Intell. Series 7, Elsevier-North Holland, New York.

R. A. Kowalski [1981], "Logic as a database language," *Proc. of Workshop on Database Theory*, Cetraro, Italy.

R. A. Kowalski [1983], "The relationahip between logic programming and logic specification," *BCS-FACS Workshop on Program Specification and Verification*, University of York.

R. A. Kowalski and D. Kuehner [1971], "Linear resolution with selector function," *Art. Intell.*, **2**, 227–260.

D. Kozen and R. Parikh [1982], "An elementary proof of the completeness of PDL," *Theor. Comp. Sci.*, **14**(1981), 113–118.

G. Kreisel [1957], "Countable functionals," Summaries of talks given at the AMS Summer Institute in Recursion Theory, 1957.

G. Kreisel [1959], "Interpretation of analysis by means of constructive functionals of finite types," in *Constructivity in Mathematics* (A. Heyting, ed.), North Holland, Amsterdam, 101–128.

S. Kripke [1963], "Semantical considerations on modal logic," *Acta Phil. Fennica*, **16**, 83–94.

S. Kripke [1972], "Naming and necessity," in *Semantics of Natural Language* (Harmon and Davidson, eds.), Reidel, 253–356.

P. J. Landin [1964], "The mechanical evaluation of expressions," *Comp. J.*, **6**, 308–320.

P. J. Landin [1965], "A correspondence between Algol 60 and Church's lambda calculus," *ACM*, **8**, 158–165.

P. J. Landin [1966a], "A lambda calculus approach," in *Advances in Programming and Non-Numerical Computation* (L. Fox, ed.), Pergamon Press, Oxford, 97–141.

P. J. Landin [1966b], "The next 700 programming languages," *Comm. ACM*, **9**, 157–164.

D. B. Lenat [1977], "Automated theory formation in mathematics," *Proc. IJCAI-77* (R. J. Reddy, ed.), 1093–1105.

C. I. Lewis and C. H. Langford [1932], *Symbolic Logic*, New York.

J. W. LLoyd [1984], *Foundations of Logic Programming*, Springer-Verlag, Berlin.

D. Loveland [1978], *Automatic Theorem Proving: A Logical Basis*, North Holland, Amsterdam.

W. Lukaszewicz [1983], "General approach to non-monotonic logics," *Proc. Eighth IJCAI*, Karlsruhe, 352–354.

J. McArthur [1979], *Tense Logic*, Reidel.

J. McCarthy [1960], "Recursive functions of symbolic expressions and their computation by machine," *Comm. Assoc. Comp. Mach.*, **3**, 184–195.

J. McCarthy [1962], "Computer programs for checking mathematical proofs," in *Recursive Function Theory* (Proc. Symp. Pure Math.), Amer. Math. Soc., Providence, Rhode Island, 219–227.

J. McCarthy [1963a], "A basis for a mathematical theory of computation," in *Computer Programming and Formal Systems* (P. Braffort and D. Hirschberg, eds.), North Holland, Amsterdam, 33–70.

J. McCarthy [1963b], "Towards a mathematical science of computation," in *Information Processing*, North Holland, Amsterdam, 21–28.

J. McCarthy [1965], "Problems in the theory of computation," *Proc. of the IFIPS Congress*, 219–222.

J. McCarthy [1980], "Circumscription as a form of non-monotonic reasoning," *Art. Intell.*, **13**, 27–39.

J. McCarthy, P. W. Abrahams, D. J. Edwards, T. P. Hart and M. I. Levin [1965], *LISP 1.5 Programmer's Manual*, MIT Press, Cambridge, Massachusetts.

D. McDermott [1982], "Non-monotonic logic II: non-monotonic modal theories," *JACM*, **29**, 33–57.

D. McDermott and J. Doyle [1980], "Non-monotonic logic I," *Art. Intell.*, **13**, 41–72.

Z. Manna [1974], *Mathematical Theory of Computation*, McGraw-Hill, New York.

Z. Manna [1980], "Logic of programs," *Proc. IFIPS Congress*, Toyko and Melbourne, 41–51.

Z. Manna, S. Ness and J. E. Vuillemin [1972], "Inductive methods for proving properties of programs," *Proc. of ACM Conference*, New York.

Z. Manna and A. Pnueli [1970], "Formalization of properties of functional programs," *J. Assoc. Comp. Mach.*, **17**, 555–569.

Z. Manna and A. Pnueli [1979], "The modal logic of programs," *Proc. Sixth Intl. Symp. on Automata, Languages, and Programming*, Graz, Austria, Lecture Notes in Computer Science 71, Springer-Verlag, Berlin, 385–409.

Z. Manna and A. Pnueli [1982], "Verification of concurrent programs: the temporal framework," in *The Correctness Problem in Computer Science* (R. S. Boyer and J. S. Moore, eds.), International Lecture Series in Computer Science, Academic Press, 215–273.

Z. Manna and R. Waldinger [1980], "A deductive approach to program synthesis," *ACM Trans. on Prog. Languages and Semantics*, **2**, 90–121.

Z. Manna and P. Wolper [1984], "Synthesis of communicating processes from temporal logic," *ACM Trans. on Programming and Systems*, **6**, 63–93.

A. Martelli and U. Montanari [1976], "Unification in linear time and space," *Internal Report B76-16*, University of Pisa.

P. Martin-Löf [1975], "An intuitionistic theory of types: predicative part," in *Logic Colloquium 1973* (H. E. Rose and J. C. Shepherdson, eds.), North Holland, Amsterdam.

P. Martin-Löf [1982], "Constructive mathematics and computer programming," in *Sixth Intl. Cong. for Logic, Phil. and Method. of Science*, North Holland, Amsterdam.

Mathlab Group [1977], *MACSYMA Reference Manual*, Technical Report, MIT.

R. E. Milne [1974], *The Formal Semantics of Computer Languages and Their Implementations*, Ph.D. Thesis, University of Cambridge.

R. E. Milne and C. A. Strachey [1976], *A Theory of Programming Language Semantics* (2 vols.), Chapman and Hall, London, and John Wiley, New York.

A. J. R. G. Milner [1973], "An approach to the semantics of parallel programs," *Proc. Convegno di Information Teorica*, Instituto di Elaborazione della Informazione, Pisa.

A. J. R. G. Milner [1975], "Processes: a mathematical model of computing agents," in *Logic Colloquium '73* (H. E. Rose and J. C. Shepherdson, eds.), 157–174.

A. J. R. G. Milner [1980], *A Calculus of Communicating Systems*, Lecture Notes in Computer Science 92, Springer-Verlag, New York.

J. Minker [1977], "Control structure of a pattern directed search system," *Technical Rept. 503*, Dept. of Computer Science, University of Maryland.

R. Moore [1983], "Semantical considerations on nonmonotonic logic," *Proc. Eighth IJCAI*, 272–279.

R. C. Moore [1985], "Semantical constructions on nonmonotonic logic," *AI J.*, **25**, 75–94.

J. H. Morris [1968], *Lambda Calculus Models of Programming Languages*, Ph.D. Diss., MIT.

B. Moszkowski and Z. Manna [1983], "Reasoning in interval temporal logic," in *Logic of Programs* (Clarke and Kozen, eds.), Lecture Notes in Computer Science 164, 371–381.

P. Naur [1969], "Proofs of algorithms by general snapshots," *BIT*, **6**, 310–316.

A. Nerode [1959], "Some Stone spaces and recursion theory," *Duke J. Math.*, **26**, 397–405.

N. Nilsson [1980], *Principles of Artificial Intelligence*, Tioga Pub. Co., Palo Alto, California.

B. Nordström [1981], "Programming in constructive set theory: some examples," *Proc. 1981 ACM Conf. on Functional Programming Languages and Computer Architecture*, 48–54.

P. Nordström and K. Petersson [1983], "Types and specification," in *Information Processing 83* (R. E. A. Mason, ed.), North Holland, Amsterdam, 915–920.

B. Nordström and J. Smith [1983], "Why type theory for programming?", *Proc., Computer Science Memo 80*, Programming Methodology Group, Dept. Computer Science, University of Göteborg.

M. J. O'Donnell [1985], *Equational Logic as a Programming Language*, MIT Press, Cambridge, Massachusetts.

S. Owicki and D. Gries [1976], "Verifying properties of parallel programs: an axiomatic approach," *CACM*, **19**, 279–284.

S. Owicki and D. Gries [1976], "An axiomatic proof technique for parallel programs," *Acta Informatica*, **6**.

R. Parikh [1978], "The completeness of propositional dynamic logic," in *Math. Found. of Computer Science 1978, Seventh Symp.*, Lecture Notes in Computer Science 64, Springer-Verlag, Berlin, 403–415.

R. Parikh [1985], *Logics of Programs* (R. Parikh, ed.), Lecture Notes in Computer Science 193.

M. S. Paterson and M. N. Wegman [1976], "Linear unification," *Proc. Eighth ACM Symp. on Theor. of Comp.*, 181–186.

L. M. Pereira and L. F. Monteiro [1981], "The semantics of parallelism and coroutining in logic programming," *Coll. on Math. Logic in Programming*, Salgotarjan, Hungary, North Holland, Amsterdam, 611–657.

G. D. Plotkin [1972], "Building-in equational theories," *Mach. Intell.*, **7** (B. Meltzer and D. Michie, eds.), 73–90.

G. D. Plotkin [1976], "A power domain construction," *SIAM J. on Comp.*, **5**, 452–487.

A. Pnueli [1977], "The temporal logic of programs," *Proc. 18th IEEE Symp. on Found. Comp.*, Providence, Rhode Island.

V. Pratt [1976], "Semantical considerations on Floyd-Hoare logic," *17th Annual IEEE Symp. on Found. of Comp. Sci.*, New York, 109–121.

V. Pratt [1980], "Applications of modal logic to programming," *Studia Logica*, **39**, 257–274.

PRL Staff [1985], *Implementing Mathematics with the Nuprl Proof Development System*, Computer Science Dept., Cornell University.

P. Raulefs, J. Siekmann, P. Szabo and E. Unvericht [1978], "A short survey of the state of the art in matching and unification problems," *AISB Quarterly*, **32**, 17–21.

R. Reiter [1978], "On closed world data bases," in Gallaire and Minker [1978].

R. Reiter [1980], "A logic for default reasoning," *Art. Intell.*, 81–132.

R. Reiter [1985], "A theory of diagnosis from first principles," *Technical Report 187/86*, Dept. of Computer Science, University of Toronto.

G. Robinson and L. Wos [1969], "Paramodulation and theorem proving in first order logic with equality," in *Machine Intelligence 4* (B. Meltzer and D. Michie, eds.), Edinburgh University Press, Edinburgh, 135–150.

J. A. Robinson [1965], "A machine-oriented logic based on the resolution principle," *J. Assoc. Comp. Mach.*, **12**, 23–41.

J. A. Robinson [1965], "Automatic deduction with hyper-resolution," *Intl. J. of Comp. Math.*, **1**, 227–234.

J. A. Robinson [1971], "Computational logic: the unification algorithm," in *Machine Intelligence 6* (B. Meltzer and D. Mitchie, eds.), Edinburgh University Press, Edinburgh, 63–72.

J. A. Robinson [1979], *Logic: Form and Function. The Mechanization of Deductive Reasoning*, Edinburgh University Press, Edinburgh.

J. A. Robinson and E. E. Siebert [1980], "Logic programming in LISP," Research Report, School and Information and Computer Science, Syracuse University, Syracuse, New York.

J. A. Robinson and L. Wos [1969], "Paramodulation and theorem proving in first order logic with equality," in *Machine Intellingence* (D. Michie, ed.), Edinburgh University Press, 103–133.

B. K. Rosen [1974], "Correctness of parallel programs: the Church-Rosser approach," *IBM Research Rpt. RC5107*, T. J. Watson Research Center, Yorktown Heights, New York.

P. Roussel [1975], "PROLOG: manuel de référence," Research Report, Art. Intell. Group of Aix-Marseille, Luminy, France.

A. Salwicki [1970], "Formalized algorithmic languages," *Bull. Acad. Polon. Sci. Math. Astron. Phy.*, **18**, 227–232.

Schröder [1890], *Vorlesungen über die Algebra der Logik* (3 vol., 1890–1905), Leipzig.

D. S. Scott [1970], "Outline of a mathematical theory of computation," *Proc. Fourth Ann. Princeton Symp. on the Semantics of Algorithmic Languages* (E. Engeler, ed.).

D. S. Scott [1970], "Constructive validity in mathematics," *Symposium on Automatic Demonstration*, Lecture Notes in Mathematics 125, Springer-Verlag, Berlin.

D. S. Scott [1972], "Lattice theory, data types, and semantics," *NYU Symp. on Formal Semantics* (R. Rustin, ed.), Prentice Hall, New York.

D. S. Scott [1973], "Models for various type-free calculi," in *Logic, Method., and Phil. of Science IV* (P. Suppes, L. Henkin, A. Joja and G. C. Moisel, eds.), North Holland, Amsterdam.

D. S. Scott [1976], "Data types as lattices," *SIAM J. on Computing*, **5**, 522–587.

D. S. Scott [1977], "Logic and programming languages," *Comm. ACM*, **20**, 634–641.

D. S. Scott [1980], "Lambda calculus: some models, some philosophy," *Kleene Symposium* (J. Barwise, K. Kunen and J. Keisler, eds.), North Holland, Amsterdam, 223–265.

D. S. Scott [1981], "Lectures on a mathematical theory of computation," *Tech. Monograph PRG-19*, Oxford University Programming Lab., Oxford, England.

D. S. Scott and G. Strachey [1971], "Towards a mathematical semantics for computer languages," *Proc. Symp. on Computers and Automata 21* (J. Fox, ed.), Polytechnic Institute of Brooklyn Press, New York.

K. Segerberg [1982], *A Completeness Theorem in the Modal Logic of Programs, Universal Algebra and Applications* (T. Traczyk, ed.), Banach Center Publications 9, PWN-Polish Scientific Publishers, Warsaw.

J. C. Shepherdson [1984], "Negation as failure," *J. Logic Programming*, **1**, 51–79.

J. C. Shepherdson [1985], "Negation as failure II," *J. Logic Programming*, 185–202.

M. Shönfinkel [1924], "Über die Bausteine der Mathematischen Logik," *Math. Ann.*, **92**, 305–316.

J. H. Siekmann [1984], "Universal unification," *Seventh Int. Conf. in Automated Deduction*, Lecture Notes in Computer Science 170, 1–42.

J. H. Siekman and G. Wrightson [1983], *The Automation of Reasoning: Collected Papers from 1957 to 1970* (2 vol.), Springer-Verlag, Berlin.

T. Skolem [1920], "Logisch-Kombinatorische Untersuchungen über die Erfüllbarkeit oder Beweisbarkeit Mathematischer Sätze nebst einer Theoreme über Dichte Mengen," in van Heijenoort.

J. Smith [1982], "The identification of propositions and types in Martin-Löf's type theory: a programming example," in *Foundations of Computer Science*, Lecture Notes in Computer Science 158 (M. Karpinski, ed.), 445–454.

S. Stenlund [1972], *Combinators, Lambda Terms and Proof-Theory*, D. Reidel, Dorddrecht.

E. Stoy [1981], *Denotational Semantics: The Scott-Strachey Approach to Programming Language Theory*, MIT Press, Cambridge, Massachusetts.

S.-A. Tärnlund [1975], "An interpreter for the programming language predicate logic," *Proc. Fourth IJCAI*, Tbilisi, Georgia, USSR, 601–608.

S.-A. Tärnlund [1977], "Horn clause computability," *BIT*, **17**, 215–226.

S.-A. Tärnlund [1978], "An axiomatic database theory," in Gallaire and Minker.

S.-A. Tärnlund [1980], *Proc. of the Logic Programming Workshop*, Debrecen, Hungary.

A. Tarski [1956], *Logic, Semantics, Metamathematics* (papers from 1923 to 1936), J. W. Woodger, tr., Oxford, Clarendon Press.

R. D. Tennent [1981], *Principles of Programming Languages*, Prentice Hall, New York.

A. S. Troelstra [1974], *Metamathematical Investigation of Intuitionistic Arithmetic and Analysis*, Lecture Notes in Mathematics 344, Springer-Verlag, Berlin.

A. M. Turing [1936], "On computable numbers, with an application to the Entscheidungsproblem," *Proc. Lond. Math. Soc.*, **42**, 230–265.

M. H. van Emden [1977], "Programming in resolution logic," in *Machine Intelligence 8*, 266–299.

K. H. van Emden [1978], "Computation and deductive information retrieval," in *Formal Description of Programming Concepts* (E. Neuhold, ed.), Proc. of Conf. on Theoretical Comp. Sci., University of Waterloo, Ontario, Canada.

K. H. van Emden and R. A. Kowalski [1976], "The semantics of predicate logic as a programming language," *J. ACM*, **23**, 733–742.

M. H. van Emden and G. J. de Lucena Filho [1982], "Predicate logic as a language for parallel programming," in Clarke and Tärnlund [1982].

J. van Heijenoort [1967], *From Frege to Gödel*, Harvard University Press, Cambridge, Massachusetts.

Moshe Vardi [1982], "The complexity of relational query languages," *14th ACM Symp. Comp. Theory*, 137–146.

J. Von Neumann [1966], *Theory of Self-Reproducing Automata*, University of Illinois Press.

C. P. Wadsworth [1976a], *Semantics and Pragmatics of the Lambda Calculus*, Ph.D. Thesis, University of Oxford.

C. P. Wadsworth [1976b], "The relation between computational semantics and denotational properties for Scott's D_ω models of the lambda calculus," *SIAM J. on Comp.*, **5**, 488–521.

D. H. D. Warren [1977], "Implementing PROLOG—compiling predicate logic programs," *DAI Research Report 39, 40*, University of Edinburgh.

D. A. Waterman [1985], *A Guide to Expert Systems*, Addison-Wesley, Reading, Massachusetts.

L. W. Weber and N. J. Nilsson [1981], *Readings in Artificial Intelligence*, Morgan Kauffman, Los Altos, California.

R. Wilensky [1984], *Lispcraft*, Norton, New York.

L. Wos, R. Overbeek, E. Lusk and J. Boyle [1984], *Automated Reasoning*, Prentice Hall, Englewood Cliffs, New Jersey.

L. Wos et al. [1984], "An overview of automated reasoning," *J. of Automated Reasoning*, **1**, 1–48.

J. J. Zeman [1973], *Modal Logic*, Oxford.

Nerode obtained his Ph.D. at the University of Chicago under Saunders MacLane in 1956, studied afterwards with Tarski and Gödel, has been at Cornell since 1959, and served as chair in mathematics in 1982–7. His mathematical logic research has been in recursive function theory, automata, recursive equivalence types, degrees of unsolvability, and the recursive content of mathematical constructions. He has had thirty-one Ph.D. students in mathematics, computer science, and applied mathematics. He has always been a consultant to industry and government in applied mathematics and computer science. These pure and applied interests have coalesced as applied logic has developed, for instance he currently holds a distinguished visiting scientist appointment with EPA to advise them on the development of expert systems and intelligent databases.

PURE AND APPLIED MATHEMATICS
FROM AN INDUSTRIAL PERSPECTIVE

H. O. POLLAK

Bell Communications Research
Morristown, New Jersey 07960

The ostensible purpose of this talk is to comment on the difference between pure and applied mathematics from the point of view of a mathematician in a high-technology, diverse, industry. A second purpose, however, must also be admitted, and that purpose is nostalgic. I shall reminisce about mathematical life in the Bell Laboratories before the divestiture of the Bell System. It was exciting and enjoyable, and the excitement and joy will probably come through in this writing. Let no one draw the conclusion that life is less exciting and less enjoyable in any of the successor companies after January, 1984; I have simply recorded, and pondered the implications of, mathematics in a structure which has, like many structures, undergone some changes.

Advanced mathematics, and mathematical research, came to the telephone business very early. Why? First, because electrical phenomena themselves are hard to perceive and to work with directly, and are therefore more likely to be modeled and studied abstractly than objects that can be immediately seen or weighed and dissected. Secondly, there is an extra complication to electical communications. The behavior of the sender of the message, the nature of the medium on which the message travels, and the message itself, are all uncertain, and can be described only in terms of probability. Who will call next, where will the call go, how long will it be before the next one after that? We cannot hope to say exactly; we can only give probability distributions for the answer. What will the noise, the loss, the phase shift, the non-linearities be on the communication path? Again, we don't know exactly, but we can know, and work with, the *probabilities* of various kinds of distortion of the signal. What will the message be? If we knew, we wouldn't have to send it! But we *can* give probability estimates of its length and frequency content. In order to handle all these uncertainties in the real world of communications, the need grew for telephone traffic theory, for the theory of random noise, for communication theory, for information theory, and for switching theory. These are among the areas of applied mathematics which owe their origin, and much of their development, to Bell Laboratories.

Mathematical scientists didn't invent, unfold, and enrich these theories by sitting at their desks, staring into space or doodling on their pads, and dreaming about what to do next. Real applied mathematics was done under constant interaction and mutual give-and-take with all the people in Bell Laboratories and the Bell System that needed sophisticated mathematical science in order to do a better job.

Mathematics and Statistics Research at Bell Laboratories had to be good in all the areas of the mathematical sciences that are important to communications—to the total research, development, and engineering effort behind communications, and to the operations of the Bell System. That included an awful lot of the mathematical sciences. But the motivation for the research came to quite a large extent not internally from mathematics, but from living in this exciting environment to the fullest, from sticking your nose into everybody's business, consulting for all comers and at very different levels of complexity. Every VP area at Bell Labs, many parts of AT&T, a number of different operating companies directly, as well as through AT&T, were sources of good problems. Part of my reason for going to the new company now called Bell Communications Research was that the operating companies continue to be a very rich source of interesting questions for research.

Please do not take this picture too far, and assume that *every* piece of research done by mathematicians and statisticians at Bell Laboratories arose from an interaction with some group or individual in Bell Labs or the Bell System. My best guess is that about half of the research efforts started this way. Problems started with someone walking in and asking for the answer to a specific mathematical or statistical question. They walked in with a specific question in some other field, which question must then be formulated in mathematical terms. They walked in, not with a specific question, but with a vague uneasiness over something they didn't understand. They didn't know what they wanted in detail, only that they wanted help. Or, mathematical scientists themselves went out and spent a day with colleagues somewhere in BTL or in the old Bell System, for the specific purpose of sticking their nose in the people's business, and seeing where they could be helpful. As I said, perhaps half of the total research effort started this way. The other half started because these mathematical scientists lived in an enormously exciting environment, read and heard about all sorts of interesting things going on, played ping-pong and ate lunch and went on canoe trips with interesting people. When you do things like this, you wake up one morning and suddenly realize that area "x" really would be an interesting area to work in, and that if you could only understand problem "y" and what was going on in it, then a particular aspect of the world of communications might indeed ultimately be improved.

I want to give you some examples, some case histories, of mathematics and statistics research at Bell Laboratories, examples that happened under my supervision, or that I have had a hand in myself. Are they typical, or are they an atrociously biased sample? Well, I have an awful lot to choose from. The 60 or so researchers in the Mathematics and Statistics Research Center of Bell Laboratories at that time, wrote about 200 papers a year—just about everything they did was published—and the total bibliography of these 60 people was about 2,000 papers and maybe 20 books. They were among the world leaders in statistics, data analysis, combinatorics, theory of algorithms, graph theory, communication theory, information theory, queuing theory, probability to name a few. So there is a lot to choose from. The examples I will cite are undoubtedly biased in the direction of having been successful. If a mathematician, at the end of a year's work, knows 200 ways which will *not* solve the following problem, that's not very interesting to talk about. But I assure you that the examples are typical of the variety of ways in which we used to work, and of the life of a mathematical scientist in one of the most exciting atmospheres for research ever created.

1. Subject: Closed Queuing Networks.

A group working on a computer communication system came for help: given the large number of nodes in the networks they were designing, the large number of different classes of customers, and the large number of customers of each class, no existing theory, and no commercially available package of computer programs, could tell them how to compute the mean waiting time for each class of customer at each node, and the variability in that waiting time. Unfortunately, it's hard to engineer the system if you can't compute these measures of its traffic-handling capability. Several mathematicians solved this problem and created a new theory of large queuing networks, by using ideas which had been developed in mathematical physics but had not previously been applied to traffic engineering. The theory was very successful, and the communication network developers are happy; the talks and published papers in the subject created quite a stir outside of BTL. The important lesson to me was: even though the mathematicians had been working in this area, no one had thought of tackling the problem of *large* networks until someone needed it and asked for help. See [1].

2. Subject: Commercial Paper.

An old friend at AT&T called to say that this time he had a *really* important problem. This was to understand and measure differences, if any, in the performance of dealers who provide commercial paper to the various Bell System companies. This was a very difficult situation to analyze: a particular issue of commercial paper was characterized by the amount of the loan, the duration of the loan, the interest rate, the state of the financial markets at the time of the loan, the particular dealer who was involved, and the particular Bell System company. There were nowhere-near enough data to analyze separately the interest rate for every combination of amount, duration, market situation, dealer, and associated company. Some clever ways of combining the data were found, and the solution was a nice application of robust smoothing and computer graphics which had been developed earlier. The result: the dealers are not all alike, far from it. The methodology developed for analyzing how good a deal was being offered became in current use in the old Bell System and, I hasten to add, by the dealers in commercial paper! It was, of course, published. See [2].

3. Subject: Simultaneous Limiting in Time and Frequency.

It is a mathematical fact that no communication signal can be both limited in time— that is, start at some time and end at some time—and also limited in the frequencies of which it is composed, that is, use no frequencies above some number of hertz. It is an engineering fact that we use such signals which are *both* "time- and band-limited," as they are called, every day, and that we get into no trouble designing systems on this basis. How come we get away with this? How can you assume something that is mathematically impossible and do successful engineering on that basis? As a result of a long series of papers, covering about 25 years, this situation is now very well understood. The theory has many applications in communications engineering; the same problem also arises in antenna theory and in optics, and the results have been very useful there as well. I personally played a major role in the first few years of this research, and I know how it started: it was in the air. Nobody came in to ask this question; but I slowly realized, from teaching an internal course in signal theory, and from talking with Claude Shannon and with many others, that this area ought to be investigated. One day, I noticed that some recent work on optimal detection of signals—a very different problem—could be used to get *started* on this question, and we were off. The influence on several different fields of engineering, and more recently on mathematics itself, has been considerable. See [3], [4], [5].

4. Subject: Cable Packing.

I was visiting a friend in that part of BTL that used to work on central office equip- ment and layout—in fact, I was a member of a committee to consider the promotion of one of his people. After the meeting, he said he had a problem he wanted to ask me about. Some central offices, particularly in big cities, have huge bundles of wires in them, many thousands of wires running from one piece of equipment to another. We know how many wires *should* be in the bundle; if the number *actually* in the bundle should happen to be much larger, this would mean that a lot of dead jumpers haven't been removed, and something needs to be done. How do you tell how many wires actually are in a big bundle? You can't count them, you can't weigh them, what do you do? One thing you can do is to take a tape measure, pull it as tight as possible around the bundle, and see what the circumference of the bundle is. By pulling the tape measure tight, you get a packing of the circular cross-sections of the wires in the bundle which is just about as tight as it can be. Mathematics says that the closest packing of equal circles covers 91% of the space, that is, only 9% of the area is in the holes between the circles. So all you have to do is to take 91% of the area surrounded by the tape measure, divide by the cross-sectional area of a single wire, and presto, you have a good estimate of the number of wires. The only trouble was, it didn't work. In experiments, it gave way too big an answer. What went wrong? The error turned out that these are not single circular wires packed as tightly as possible, they are twisted pairs! And twisted pairs can't pack together anywhere near as tightly as single wires. How tightly? This was a new problem; nobody knew. Ed Gilbert showed that the tightest possible lattice packing of twisted pairs covers only about 57% of the cross-sectional area; this gave a reasonable result in the real world. Really deep new mathematics? Probably not. But a useful new result, and a pretty piece of mathematical research. See [6].

5. Quantization and the Gaussian Channel.

The packing of circles in the last example has its analogs in higher dimensions. The mathematics in this example requires packing spheres in 8, or 16, or even more dimensions; thinking of stacked cannon balls in a 3-dimensional Civil War monument will help visualize what's coming. If one of a collection of signals, all limited in power, is sent over a channel in which it is corrupted by noise, then the noise can occasionally be so bad that the receiver will be misled into thinking a different signal has been sent. That's an error, and you have to keep down the error rate in communication. According to Shannon's mathematical theory of communication, you can improve the error rate, at the price of delay and extra complexity, by collecting m consecutive signals you want to send, and representing them by lattice points in m dimensions each of which is the center of one of our "cannon balls." Get the tightest packing of these cannon balls, and you've got a lower error rate. Many mathematicians and electrical engineers have been working on this fundamental problem for many years. In the last five years, two very exciting things happened: the problem of quantizing signals, for example for speech encoding and transmission, turned out to be the same problem; and recent advances in lattice packing in high dimensions, and in finite group theory, have led to real improvements in the best packings. Furthermore, applications of these new lattice packings to the Gaussian channel have made a real difference to the achievable error rates in the transmission of data over ordinary telephone lines. This aspect of the problem began at the specific request of the data transmission people, the results are also being used by speech researchers, and it was all made possible by recent advances in what most mathematicians would have considered a pretty pure subject—high-dimensional lattice packings. See [7].

6. Subject: Engineering Effectiveness.

A man from the engineering area at the former AT&T called because he had a tough problem: he wanted to develop an index for measuring the effectiveness of district engineers in the operating companies, an index which struck a proper balance between motivation towards cost-cutting and motivation towards giving good service. A statistical study was made of his data, and it was found that a single index really couldn't convey the information he wanted to get across; several levels of indices were necessary. One difficulty, for example, is that engineers for rural and for urban districts simply cannot be measured in exactly the same way. These several levels of indices allow each engineer to examine and improve several distinct components of good performance. As a result of this work, EDEM (Effectiveness of District Engineers Measurement Plan) was developed. On the statistical side, several new clustering display techniques needed to be invented, and were duly published in the statistical literature. See [8].

7. Atmospheric Pollution.

At the specific request of W. O. Baker, then president of Bell Laboratories, several statisticians attended a meeting with officials of the New Jersey Department of Environmental Protection to discuss some new proposed federal EPA rules governing automotive emissions. The former Bell System, as the largest non-governmental operator of vehicles, has a vital interest in regulations governing hydrocarbon emissions. It was therefore very sensible for chemists and statisticians at Bell Laboraboriess to join forces and make every effort to understand the fundamental science behind atmospheric pollution. Data on atmospheric constituents and on meteorological conditions were made available for many locations in the Northeast, and it was possible to undertake a major scientific investigation of what these data really showed. Some sample results: in the Northeast, hydrocarbon emissions were *not*, as had previously been believed, the major determinant of ozone pollution. Ozone on Sunday morning, when hydrocarbon emissions are low, was no lower than during weekday morning rush hours. The same is true of a variety of other products of smog chemistry. Ozone levels *were* greatly affected by solar radiation and by wind speed. While hydrocarbon effects were minimal, various oxides of nitrogen had a much greater effect on smog chemistry than had previously been supposed. This work, of considerable national significance, required the close collaboration of researchers in chemistry and in statistics, and of various governmental agencies. It was all published, and led to the modification of the proposed standards. It was also a rich source of fundamental statistical research problems, and led for example to the first satisfying solution to the problem of how to smooth scatter plots. See [9].

8. Absenteeism.

A friend at AT&T Human Factors called and asked some questions about tests of significance and factor analysis. The researchers, in turn, as any applied mathematician and statistician soon learns to do, asked why he wanted to know. The problem, it turned out, dealt with understanding the absenteeism pattern of operators for New England Tel & Tel, and thereby finding a better management strategy. The difficulty of the problem is one with which the then existing statistical theory could not deal adequately. The data on absence intervals of operators forms a length-biased sample, since operators absent when the data gathering starts, and those absent when it ends, distort the information on how long absences are, and how they relate to other personal characteristics. In the data analysis, it turns out that the distribution of lengths of absences is well described as a mixture of two exponential distributions, which one might call disability and incidental absences. There is also a change in parameters at about age 35. Some very interesting new statistical theory has been developed as a result of this study, the research is continuing, and in fact there was a doctoral dissertation at the University of Michigan for one of the statisticians involved, on this new theory. See [10].

9. INOS.

One of the mathematicians in the organization had become, as a result of a number of interesting problems over the years, an expert in heuristic programming, that is in the approximate solution on the computer of combinatorial problems so huge and messy that there was no hope of a totally analytic treatment. Some years ago, as a result of some friends at what was then Long Lines Marketing, he became interested in the problem of preparing, for the benefit of large corporate customers of Long Lines, a plan for the mix of communications services that would meet the company's projected needs at minimum cost. Together with a research computer scientist at Bell Laboratories, this mathematical researcher developed the methods and computer programs for INOS, the Interactive Network Optimization System, which became a standard tool, both for Long Lines Marketing and for the planning behind new tariffs and service offerings at AT&T. This direct path from mathematical research on combinatorial algorithms to the marketing of telecommunications services is an exciting and satisfying experience and has led to a new methodology, published and more widely applicable to similar problems. See [11].

10. Charge Coupled Devices.

On another occasion some years ago, a device physicist from one of the semi-conductor device development groups came to seek help in solving an equation. As is often the case, the equation he wanted to solve was the wrong one; his mathematical formulation of the problem was incorrect. Working with him, a mathematician formulated the problem correctly, at which point it became clear this was an extremely interesting and important problem.

From this beginning there developed an important series of research efforts. Out of this work came very interesting mathematics and an exciting new technique in numerical computation. This work made an important contribution to understanding the operation of charge coupled devices, which were invented at the old Bell Laboratories. Detailed accounts of this research were published in scholarly journals. As is sometimes the case, charge coupled devices have been much more extensively applied outside the U.S. telephone system than inside. Consequently this particular mathematical research has had its greatest impact outside the Bell System. See [12], [13].

What do we learn from these examples, and from hundreds of others in my experience as director of mathematics and statistics research? First of all, mathematicians and statisticians at Bell Laboratories did an enormous amount of research, leading to a couple of hundred published papers every year. But the published papers are far from the full story. A paper hardly ever tells *why* the research was undertaken in the first place, what led to this particular series of questions, and what the results mean to the outside (or mathematics) world.

It is simply not the fashion to write this information down; furthermore, journal editors, hard pressed for space, wouldn't allow you to publish it even if you did record it. When a situation arises where knowing the origins and destinations of research is important, you have to ask those who performed, or supervised the ones who performed, the research.

In my opinion, a critical aspect of the difference between pure and applied mathematics is the precise reason *why* the research was undertaken. This is rarely documented in a mathematical paper, and it is for this reason that it has formed the bulk of this presentation. If the work was undertaken for the inherent mathematical interest of the problem, it is an example of pure mathematics. If it was undertaken because of the light it was hoped to shed on some other part of mathematics, it is also usually defined as pure mathematics. If the work was undertaken because of the light it was hoped to shed on some other field of human endeavor, or on some aspect of real life, then I define it as applied mathematics. Thus I feel I cannot classify an individual theorem as either pure or applied—it depends on why you are interested.

Let us dig a little deeper. What about a sequence of theorems—that is, a theory? Suppose you start from some external situation, make a mathematical model and succeed in proving a theorem. The real question is, "What do you do next?" One thing that may happen is that the theorem opens up a new series of questions in mathematics itself—generalizations, parallels in similar structures, etc. On the other hand, you may take the theorem, see what it says about the original situation outside of mathematics from which you started, and decide what you want to understand next in the outside world—now that you know this. If the sequence is governed by the logic of the mathematical structure itself, then you are acting as a pure mathematician. If it is governed by the logic of the application, then you are acting as an applied mathematician.

A third aspect of the difference between pure and applied mathematics from an industrial point of view is the matter of *stability*. A useful applicable theorem must be stable, that is, the result must change only a little bit if the assumptions change only a little bit. As Richard W. Hamming, a controversial mathematician, computer scientist, and numerical analyst, retired from Bell Laboratories and now teaching at the Naval Postgraduate School, said: "If the design of this airplane depends on the difference between the Riemann and the Lebesgue integral, I don't want to fly it." More seriously, in a theory like Fourier transforms there is a wealth of both stable and unstable theorems. When you apply Fourier transforms to electrical engineering or optics or antennas, the result had better be stable since the mathematical model can never be perfectly accurate.

References.

[1] J. McKenna, D. Mitra and K. G. Ramakrishnan, "A class of closed Markovian queuing networks: integral representations, asymptotic expansions, generalizations," *Bell Sys. Tech. J.*, **60** (May–June 1981).

[2] A. Cohen, J. M. Schilling and I. J. Terpenning, "Dealer issued commercial paper: analysis of data," *Proceedings of the Business and Economic Statistics Section, American Statistical Association*, 1979, 162–164.

[3] H. O. Pollak and D. Slepian, "Prolate spheroidal wave functions, Fourier analysis and uncertainty – I," *Bell Sys. Tech. J.*, **40** (January 1961), 43–64; Monograph 3746.

[4] H. O. Pollak and H. J. Landau, "Prolate spheroidal wave functions, Fourier analysis and uncertainty, II," *Bell Sys. Tech. J.* (January 1961).

[5] H. J. Landau and H. O. Pollak, "Prolate spheroidal functions, Fourier analysis and uncertainty, III. The dimension of the space of essentially time- and band-limited signals," *Bell Sys. Tech. J.*, **41** (July 1962), 1295–1336.

[6] E. N. Gilbert, "The packing problem for twisted pairs," *Bell Sys. Tech. J.*, **58** (December 1979), 2143–2162.

[7] J. H. Conway and N. J. A. Sloane, "Fast quantizing and decoding algorithms for lattice quantizers and codes," *IEEE Trans. on Information Theory*, **IT-28** (1982), 227–232.

[8] E. B. Fowlkes, J. D. Gabbe and J. E. McRae, "A graphical technique for making a two-dimensional display of multidimensional clusters," *Proceedings of the Business and Economics Statistics Section of the American Statistical Association*, 1976.

[9] W. S. Cleveland and T. E. Graedel, "Photochemical air pollution in the Northeast United States," *Science*, **204** (1979), 1273–1278.

[10] Y. Vardi, "Absenteeism of operators: a statistical study with managerial applications," *Bell Sys. Tech. J.*, **60** (1981), 13–38.

174

[11] S. Lin, "Effective use of heuristic algorithm in network design," in *The Mathematics of Networks*, AMS Short Course Publication Series, 1982.

[12] J. McKenna and N. L. Schryer, "On the accuracy of the depletion layer approximation for charge-coupled devices," *Bell Sys. Tech. J.*, **51** (1972), 1471–1485.

[13] J. McKenna and N. L. Schryer, "The potential in a charge-coupled device with no mobile minority carriers," *Bell Sys. Tech. J.*, **52** (December 1973), 1765–1793.

After 32 years at Bell Laboratories, Pollak joined Bell Communications Research as Assistant Vice-President for Mathematical, Communications, and Computer Sciences Research when Bellcore was formed as a result of the Bell System divestiture. Pollak and Young first met on the Committee on the Undergraduate Program of the Mathematical Association of America around 1960; Pollak always maintained an interest in education along with his research in the mathematics of the communications industry. Among their joint activities, a trip to Africa in 1968 in order to evaluate recent activities in mathematics education was particularly noteworthy.

LETTER VALUES IN MULTIVARIATE
EXPLORATORY DATA ANALYSIS

DONALD ST. P. RICHARDS

Department of Statistics
University of North Carolina
Chapel Hill, North Carolina 27514

RAMESHWAR D. GUPTA

Division of Mathematics,
Computer Science and Engineering
University of New Brunswick
St. John, New Brunswick, Canada

ABSTRACT

By projecting multivariate data onto hyperplanes, letter values are used to construct a statistic for testing the hypothesis that a multivariate data set follows a completely specified distribution. Further examples of multivariate distributions whose projections have nicely spaced letter values are constructed.

1. Introduction.

In classical statistical analysis, great emphasis is placed on the behavior of statistical procedures when the size of the underlying sample is large. Thus, large-sample properties such as consistency, asymptotic normality and asymptotic relative efficiency play a prominent role in classical work. We refer to [1] for an introductory treatment of these concepts.

Recently, exploratory (or empirical) methods for analyzing data sets have gained popularity and importance in statistical work. There are a number of reasons for the increased attention being paid to the new methods, and we mention two of the most important. First, classical procedures often assume the existence of a large sample size. However, data sets are usually small and their statistical properties are frequently more complicated than in the case of large samples. Secondly, in exploratory statistical analysis, many of the simple statistics used to summarize information provided by a data set are computed using sorting and counting rules. Not only are these summaries of the "quick and dirty" genre, but they are also quite *robust* [6];

175

that is, large variations in a small portion of the data set cause only small changes in the values of the summary statistic. In general, exploratory statistics are particularly useful when the investigator needs to quickly determine important information about data sets, such as the presence of outliers, or say, whether the data set was drawn from a Gaussian population.

In another direction, classical summaries such as the sample mean and sample variance, while highly nonrobust, can be used to identify the probabilistic distribution underlying a random sample. The book [8] presents a thorough treatment of the characterization of probability distributions through properties of classical summary statistics.

Similar results for exploratory summaries are rare. In [5], it was recently shown how one class of exploratory statistics, the *letter values*, could be used to identify the underlying probability distribution. As an application of those results, [5] presented a new method for testing the hypothesis that a data set is drawn from a given, completely specified population, e.g., the standard Gaussian population.

In this paper, we extend ideas of [5] to the analysis of multivariate data. In particular, we use letter values to construct a statistic for testing the hypothesis that a sample of multivariate data is drawn from a uniquely specified population. The basic idea is to reduce the multivariate data to univariate observations by applying projections, and then to analyze the univariate data using the results of [5]. It should be noted that the procedure of projecting multivariate data onto hyperplanes has been successfully applied elsewhere [2], [7].

2. Letter Values.

Suppose that $F(x)$, $x \in R$, is an absolutely continuous distribution function with continuous probability density $f(x) = F'(x)$. The kth **letter value** is $x_k = F^{-1}(a_k)$ where

$$a_k = \begin{cases} 2^{k-1}, & k = 0, -1, -2, \ldots \\ 1 - 2^{-(k+1)}, & k = 1, 2, 3, \ldots \end{cases} \tag{1}$$

Thus, x_0 is the median, x_1 is called the upper fourth, x_2 the upper eighth, etc. Similarly, x_{-1} is called the lower fourth, x_{-2} the lower eighth, etc. The terminology "letter values" is due to Tukey [12] who denoted x_0 by M (for median), $x_{\pm 1}$ by F (for fourths), $x_{\pm 2}$ by E (for eighths), etc. Efficient methods for computing the letter values of a data set are provided in [6], [12].

3. Nicely Spaced Letter Values.

Let X be a random variable which is uniformly distributed on the interval $(-1, 1)$. Then it is easily shown through (1) that the corresponding letter values x_k are such that $x_{-k} = -x_k$, $k \geq 0$, and the **spacings** $x_{k+1} - x_k$ between successive upper letter values form a geometric progression with common ratio $r = 1/2$. Another probability distribution with nicely spaced letter values (cf. [6; p. 56]) is the Laplace distribution whose density function is

$$f(x) = (2\sigma)^{-1} \exp(-|x|/\sigma), \qquad -\infty < x < \infty, \tag{2}$$

where $\sigma > 0$. Here, $x_{-k} = -x_k$ for all k; and the spacings $x_{k+1} - x_k$ are constant. Motivated by these two examples, the following result was recently established in [5].

Theorem 1 ([5]). *Suppose that X is a random variable with density function $f(x)$. Assume that $f(x)$ is (i) continuous and symmetric about the median x_0, (ii) differentiable everywhere on the range of X, except possibly at x_0, (iii) log-concave and monotonic decreasing for $x > x_0$. If the spacings between successive letter values form a geometric progression,*

$$x_{k+1} - x_k = ar^k, \qquad k = 0, 1, 2, \ldots \tag{3}$$

where $0 < r < 1$, $a > 0$, then up to location and scale, $f(x) = f_\beta(x)$ with

$$f_\beta(x) = \begin{cases} \frac{1}{2}\beta(1 - |x|)^{\beta-1}, & |x| < 1 \\ 0 & , \quad elsewhere, \end{cases} \tag{4}$$

where $r = 2^{-1/\beta}$, $1 < \beta < \infty$.

If the density function $f(x)$ is assumed to be log-convex and monotonic increasing then we again get $f = f_\beta$, but with $0 < \beta < 1$. Using limiting arguments, or by adopting the hypothesis that $f(x)$ is log-concave and monotonic decreasing, we can also treat the case $r = 1$ when $f(x)$ is the Laplace density function (2).

When $r > 1$ in (3), so that $x_k \to \infty$ as $k \to \infty$, it is shown in [5] that a result analogous to Theorem 1 holds and characterizes the family of density functions

$$g_\beta(x) = \frac{\beta}{2(1 + |x|)^{\beta+1}}, \qquad -\infty < x < \infty, \tag{5}$$

where $\beta > 0$.

If the random variable X has the density function $f_\beta(x)$ where $1 \leq \beta < \infty$ and Y has the density $g_\beta(x)$, then it is noted in [5] that

$$1 + X \stackrel{\mathcal{L}}{=} \frac{1}{1 + Y}; \tag{6}$$

that is, the variables $1 + X$ and $(1 + Y)^{-1}$ have the same probability distribution. We note also that if the random variable Z has the Laplace density function (2), then Y has a **mixture representation**

$$Y \stackrel{\mathcal{L}}{=} ZW^{-1} \tag{7}$$

where W is independent of Z, and W has a gamma distribution with density function

$$h(x) = \frac{1}{\Gamma(\beta)} e^{-x} x^{\beta-1}, \qquad x > 0.$$

One very important case covered by Theorem 1 is the uniform distribution, arising when $\beta = 1$. For an application, consider the problem of testing the hypothesis H_0 that a random sample $X_1, X_2, \ldots, X_n$ is drawn from a continuous population whose distribution function $F(x)$ is completely specified; especially important is the case when $F(x) = \Phi(x)$, where

$$\Phi(x) = (2\pi)^{-1/2} \int_{-\infty}^{x} e^{-t^2/2} \, dt \tag{8}$$

is the standard Gaussian distribution function, which arises in many contexts [6; p. 265].

For the hypothesis H_0 to be valid, a well known necessary and sufficient condition is that the transformed data $Y_1 = F(X_1), \ldots, Y_n = F(X_n)$ follow the uniform distribution on $(0, 1)$. Assuming that the true distribution of the Y_i's satisfies the assumptions of Theorem 1, then the hypothesis H_0 is valid if and only if the letter values $y_1, y_2, \ldots$ of the transformed data form a geometric progression (4) with $r = 1/2$. Thus, the problem of testing for the distribution function $F(x)$ has been reduced to the simpler task of testing for uniformity. In [5], we used the letter values $y_1, y_2, \ldots$ to form the test statistic $R = S(y_{-N}, \ldots, y_N)$ where

$$\begin{aligned} S(y_{-N}, \ldots, y_N) = \sum_{k=-N}^{-1} \left[\frac{y_k - y_{k-1}}{y_{k+1} - y_k} - \frac{1}{2} \right]^2 + \left[\frac{y_1 - y_0}{y_0 - y_{-1}} + 1 \right]^2 \\ + \sum_{k=1}^{N} \left[\frac{y_{k+1} - y_k}{y_k - y_{k-1}} - \frac{1}{2} \right]^2, \end{aligned} \tag{9}$$

where N is to be determined by the investigator. Large values of R suggest that the y_k's are not the letter values of a uniform distribution, hence that H_0 cannot be accepted.

Roughly speaking, the integer $N > 1$ should be chosen to equal the smallest value of $|k|$ for which the term in square brackets in (9) is negligible. For a wide class of distributions, there is nothing to be gained by choosing extremely large values for N; indeed if $\bar{F}(x) = 1 - F(x)$, and $\bar{F}^{-1}$ exists, and has a nonzero derivative at $x = 0$, then L'Hôpital's rule and (1) will show that

$$\lim_{k \to \infty} \frac{y_{k+1} - y_k}{y_k - y_{k-1}} = \frac{1}{2}. \tag{10}$$

Then, a suitable value for N is clearly dependent on the tail behavior of $F(x)$.

The distribution of the test statistic R, when the hypothesis H_0 is valid, is quite complicated. Consequently, simulation studies of R are now being performed and will appear elsewhere.

4. Multivariate Distributions.

In the case of multivariate random variables, the notion of letter value defined by (2) does not make sense since multivariate cumulative distribution functions are not invertible. In order to analyze multivariate data using letter values, we shall need some definitions. For the rest of the paper, it is assumed that X is an n-dimensional, absolutely continuous random vector with distribution function $F(x)$ and continuous density function $f(x)$. We use $\langle \cdot, \cdot \rangle$ and $\|\cdot\|$ respectively for the standard inner product and norm on n-dimensional Euclidean space $\mathbb{R}^n$.

Let ω be a unit vector. The kth **upper letter value** $x_k(\omega)$ **of the random vector** X **in the direction** ω is the kth upper letter value of the random variable $\langle \omega, X \rangle$. The **lower letter value** $x_{-k}(\omega)$ is defined similarly. Thus, we use the (one-dimensional) **projections** $\langle \omega, X \rangle$ of X to define letter values for X.

Let us now construct a multivariate distribution whose letter values $x_k(\omega)$ are equally spaced and symmetric, viz.

$$x_{k+1}(\omega) - x_k(\omega) = \text{const}, \tag{11}$$

$$x_{-k}(\omega) = -x_k(\omega), \tag{12}$$

for all $k = 0, 1, 2, \ldots$ and all $\|\omega\| = 1$. Let the random vector be centered at zero, with positive-definite covariance matrix Σ. Then $\langle \omega, X \rangle$ has mean zero and variance $\langle \omega, \Sigma \omega \rangle$. If we assume that the variable $Y = \langle \omega, X \rangle$ has the Laplace density function

$$g(y) = (2\sigma)^{-1} \exp(-|y|/\sigma), \qquad -\infty < y < \infty$$

where $0 < \sigma^2 = \text{Var}(Y) = \langle \omega, \Sigma \omega \rangle$, then by some remarks in Section 3, Y has equally spaced letter values. Further the characteristic function of Y is

$$\hat{g}(t) := E(e^{itY}) = \int_{-\infty}^{\infty} e^{ity} g(y)\, dy$$
$$= (1 + t^2 \sigma^2)^{-1},$$

where the last equality follows from [4; p. 476]. In particular,

$$\hat{g}(1) = E(e^{i\langle \omega, X \rangle}) = \frac{1}{1 + \langle \omega, \Sigma \omega \rangle} \tag{13}$$

which is the characteristic function of X. Since (13) is integrable w.r.t. ω over $\mathbb{R}^n$ then X possesses a density function given through Fourier inversion by

$$f(x) = (2\pi)^{-n} \int_{\mathbb{R}^n} \frac{e^{-i\langle \omega, x \rangle}\, d\omega}{1 + \langle \omega, \Sigma \omega \rangle} \tag{14}$$

As we now show,

$$f(x) = (2\pi)^{-n/2} (\det \Sigma)^{-1/2} \left\| \Sigma^{-1/2} x \right\|^{-m} K_m \left(\left\| \Sigma^{-1/2} x \right\| \right), \qquad x \neq 0, \qquad (15)$$

where $m = (n-2)/2$ and $K_m(\cdot)$ is the modified Bessel function of the third kind [13]. In (14), replace ω by $\Sigma^{-1/2}\omega$ where $\Sigma^{-1/2}$ is the unique positive definite square root of Σ^{-1}. Then we have $f(x) = (2\pi)^{-n} (\det \Sigma)^{-1/2} f_0(\Sigma^{-1/2} x)$, where

$$f_0(x) = f_0(-x) = \int_{\mathbb{R}^n} \left(1 + \|\omega\|^2\right)^{-1} e^{i\langle \omega, x \rangle} \, d\omega.$$

Introducing polar coordinates: $\omega = ru$, $r > 0$, $\|u\| = 1$; we obtain

$$f_0(x) = \int_0^\infty \left(1 + r^2\right)^{-1} r^{n-1} \int_{\|u\|=1} e^{ir\langle u, x \rangle} \, du \, dr$$

where du is surface measure on the unit sphere in $\mathbb{R}^n$. By [13],

$$\int_{\|u\|=1} e^{i\langle u, x \rangle} \, du = (2\pi)^{n/2} \|x\|^{-m} J_m(\|x\|), \qquad x \neq 0, \qquad (16)$$

where J_m is the classical Bessel function, so that

$$f_0(x) = (2\pi)^{-n/2} \|x\|^{-m} \int_0^\infty r^{n/2} (1 + r^2)^{-1} J_m(r\|x\|) \, dr, \qquad x \neq 0. \qquad (17)$$

Then (15) follows from (16) by way of an integral in [13; Section 13.6, (2)]. Thus, the random vector X having the density function (15) has equally spaced letter values in every direction. The density function (15) has also arisen in other contexts [9].

It is more difficult to construct random vectors whose letter values $x_k(\omega)$ are spaced in geometric progression,

$$\begin{aligned} x_{k+1}(\omega) - x_k(\omega) &= ar^k, \qquad k = 0, 1, 2, \ldots \\ x_{-k}(\omega) &= -x_k(\omega) \end{aligned} \qquad (18)$$

where both a and r may depend on ω. First let us attempt to construct a vector X in $\mathbb{R}^n$ such that $Y = \langle \omega, X \rangle$ is uniformly distributed on $\left(-\|\omega\|, +\|\omega\|\right)$ for all $\omega \neq 0$. The existence of such a vector X falls within the framework of results of Eaton [3]; in particular, [3] shows that X can have at most four components, i.e., $n \leq 4$. Since $Y = \langle \omega, X \rangle$ is uniformly distributed on $\left(-\|\omega\|, +\|\omega\|\right)$, then the characteristic function of Y is

$$\begin{aligned} \phi_Y(t) = E(e^{itY}) &= \frac{1}{2\|\omega\|} \int_{-\|\omega\|}^{\|\omega\|} e^{ity} \, dy \\ &= \frac{\sin t\|\omega\|}{t\|\omega\|}, \qquad \omega \neq 0. \end{aligned}$$

Setting $t = 1$, we have

$$\phi_X(\omega) = \left(\sin \|\omega\|\right)/\|\omega\|, \qquad \omega \neq 0,$$

the characteristic function of X. When $n = 2$, we can find an explicit formula for the density function of X; in this case, Fourier inversion shows that the density function is

$$f(x) = (2\pi)^{-1} \int_0^\infty J_0\big(r\|x\|\big)(\sin r)\, dr$$
$$= \begin{cases} (2\pi)^{-1}\big(1 - \|x\|^2\big)^{-1/2}, & 0 < \|x\| < 1 \\ 0, & \text{otherwise.} \end{cases} \tag{19}$$

Therefore the letter values $x_k(\omega)$ of the vector X satisfy (18) with $r = 1/2$. We note also that it is far more difficult to construct distributions whose letter values satisfy (18) for arbitrary $r \in (0,1)$. In particular, the results of Eaton [3] show that such distributions can only exist for $n \leq 5$.

Finally, we present a generalization of Theorem 1 as it pertains to the Laplace distribution (3) and its multivariate analog (15). Recall [3] that a random vector X in $\mathbb{R}^n$ is **isotropic** if $P(X = 0) = 0$ and if X has the same probability distribution as HX for any $n \times n$ orthogonal matrix H.

Theorem 2. *Let X be an n-dimensional absolutely continuous, isotropic random vector with density function $f(x) = g\big(\|x\|^2\big)$. If $g(t)$ is log-concave, $t^{n-1}g(t)$ is monotone decreasing for $t > 0$, and the letter values $x_k(\omega)$ of X satisfy (11), then X has the generalized Laplace density function (15).*

Proof. Since X is isotropic, then $Y = \langle \omega, X \rangle$ has a density function $h(y)$ which is independent of ω and is related [3], [10] to $g(t)$ by

$$h(-y) = h(y) = c \int_y^\infty t^{n-2}g(t)\left(1 - \frac{y^2}{t^2}\right)^{m-1/2} dt$$
$$= cy^{n-1} \int_1^\infty t(t^2 - 1)^{m-1/2}g(ty)\, dt, \qquad y > 0, \tag{20}$$

where c is an appropriate normalizing constant. Since $g(y)$ is log-concave then (20) shows that $h(y)$ is also log-concave for $y > 0$. Further, $h(y)$ is monotonic decreasing along with $y^{n-1}g(y)$. Hence, our remarks above (Theorem 1, infra) concerning the Laplace distribution show that $Y = \langle \omega, X \rangle$ has a Laplace distribution (3). By the Cramér-Wold theorem, the distribution of X is uniquely determined, and via our construction above, X has the density function (15).

5. Testing for a Completely Specified Distribution.

Consider the problem of testing the hypothesis H_0 that a random sample $X_1, X_2, \ldots,$ X_m is drawn from a continuous multivariate population with completely specified distribution function $F(x)$. As in [5] we construct a statistic, based only on letter values, for testing H_0. For $\|\omega\| = 1$, let $Y_i = \langle \omega, X_i \rangle$, $i = 1, 2, \ldots, m$; since $F(x)$ is known, then so is the distribution function $F_\omega(y)$ of the Y_i. In particular, H_0 is valid if and only if for all ω, the transformed variables $W_1 = F_\omega(Y_1), \ldots, W_m = F_\omega(Y_m)$

follow the uniform distribution on $(0, 1)$. Then, the spacings $x_{k+1}(\omega) - x_k(\omega)$ of the letter values of the W_i form a geometric progression with common ratio $r = 1/2$. For testing H_0, we propose the statistic

$$R = \sup_{\|\omega\|=1} S\big(x_{-N}(\omega), \ldots, x_N(\omega)\big) \tag{23}$$

where $N = N(\omega)$ is to be chosen by the investigator. If R is large then for some ω the spacings $x_{k+1}(\omega) - x_k(\omega)$ do not form a geometric progression with ratio $r = 1/2$, suggesting that the corresponding W_i are not uniformly distributed. Therefore, H_0 is not to be accepted for sufficiently large values of R.

6. Concluding Remarks.

Central to our results is the Cramér-Wold device: the distribution of an n-dimensional random vector X is uniquely determined by the distribution of the one-dimensional projections $\langle \omega, X \rangle$, $\omega \neq 0$. Further, we may even restrict ω to satisfy $\|\omega\| = 1$. This method of linearizing multivariate data analysis is currently enjoying wide popularity. Especially, *projection pursuit* procedures have been shown [2], [7] to provide advantages over some classical procedures.

The idea of analyzing a function $f(x)$ on $\mathbb{R}^n$ through its integrals $f_\omega(y)$ over $(n - 1)$-dimensional hyperplanes $\langle \omega, x \rangle = y$ is the classical Radon transform [11]. It is known [11, Theorem 4.1] that if $f(x)$ is square-integrable with compact support, then $f(x)$ is uniquely determined by $\{f_\omega(y)\}$ where ω runs over any infinite subset of the unit sphere S^{n-1}. This proves that, in (23), it is sufficient to take the supremum over infinite subsets of S^{n-1} whenever the appropriate density function has compact support.

Acknowledgment. The authors were partially supported by grants from the National Science Foundation (MCS-8403381), the Research Council of the University of North Carolina, and by the National Scientific and Engineering Research Council, grant no. A-4850.

References.

[1] P. J. Bickel and K. A. Doksum, *Mathematics Statistics*, Holden-Day, California, 1977.

[2] P. Diaconis and D. A. Freedman, "Asymptotics of graphical projection pursuit," *Annals of Statist.*, **12**(1984), 793–815.

[3] M. L. Eaton, "On the projections of isotropic distributions," *Annals of Statist.*, **9**(1981), 391–400.

[4] W. Feller, *An Introduction to Probability Theory and Its Applications*, Vol. II, Wiley, New York, 1966.

[5] R. D. Gupta and D. St. P. Richards, "When is a distribution determined by its letter values?", Univ. of North Carolina Mimeo Series, no. 1588, 1985.

[6] D. C. Hoaglin, F. Mosteller and J. W. Tukey (eds.), *Understanding Robust and Exploratory Data Analysis*, Wiley, New York, 1983.

[7] P. J. Huber, "Projection pursuit," *Annals of Statist.*, **13**(1985), 435–475.

[8] A. M. Kagan, Y. Linnik and C. R. Rao, *Characterization Problems in Mathematical Statistics*, Wiley, New York, 1973.

[9] D. St. P. Richards, "Hyperspherical models, fractional derivatives and exponential distributions on matrix spaces," *Sankhyā*, Ser. A, **46**(1984), 155–165.

[10] D. St. P. Richards, "Positive definite symmetric functions on finite-dimensional spaces, I," *J. Multivariate Analysis*, **19**(1986).

[11] K. T. Smith, D. C. Solmon and S. L. Wagner, "Practical and mathematical aspects of the problem of reconstructing objects from radiographs," *Bull.*, *A.M.S.*, **83**(1977), 1227–1270.

[12] J. W. Tukey, *Exploratory Data Analysis*, Addison-Wesley, Reading, 1977.

[13] G. N. Watson, *A Treatise on the Theory of Bessel Functions*, 2nd ed., Cambridge University Press, New York, 1944.

Donald Richards did his graduate work at the University of the West Indies (UWI), in Kingston, Jamaica. He has taught at UWI, at the University of North Carolina at Chapel Hill, and at the University of Wyoming (on leave from North Carolina). His principal fields of interest are generalized special functions and multivariate statistics.

Rameshwar Gupta did his graduate work at Dalhousie University, in Halifax, Nova Scotia. He has taught at the University of the West Indies, at the University of Saskatchewan, and at the University of New Brunswick. His primary interests are in the fields of multivariate and applied statistics.

NEWTON'S METHOD ESTIMATES
FROM DATA AT ONE POINT

STEVE SMALE

Department of Mathematics
University of California, Berkeley
Berkeley, California 94720

In honor of Gail Young

Newton's method and its modifications have long played a central role in finding solutions of non-linear equations and systems. The work of Kantorovich has been seminal in extending and codifying Newton's method. Kantorovich's approach, which dominates the literature in this area, has these features: (a) weak differentiability hypotheses are made on the system, e.g., the map is C^2 on some domain in a Banach space; (b) derivative bounds are supposed to exist over the whole of this domain. In contrast, here strong hypotheses on differentiability are made; analyticity is assumed. On the other hand, we deduce consequences from data at a single point. This point of view has valuable features for computation and its theory. Theorems similar to ours could probably be deduced with the Kantorovich theory as a starting point; however, we have found it useful to start afresh.

The results in this paper are important for our construction of global algorithms based on Newton's method, and for estimation of the efficiency of those algorithms. The idea is simply to apply the theorems here to a finite sequence of equations of the form $f(z) - t_i f(z_0) = 0$, $0 \leq t_i \leq 1$, to solve $f(z) = 0$. That development will be exposed in a subsequent article.

The classical work can be found in Kantorovich and Akilov (1964), as well as Ostrowski (1973). More recent work in this direction includes Gragg and Tapia (1974), Rall (1974), and Traub and Wozniakowski (1979). Some of the spirit here is reflected in Smale (1981, 1985). Important for me have been conversations with Mike Shub and our two joint papers (1985, 1986). Trying to understand the important work of Renegar [8] was a strong motivation for this paper. Ideas of Kim (1985) are closely related also.

Section 1.

To indroduce the ideas, we give some of the results first for the simple case of a single polynomial. Consider a complex polynomial $f(z) = \sum_0^d a_i z^i$. To solve $f(\varsigma) = 0$, following Newton's method, let $z_0 \in \mathbb{C}$ and inductively $z_k = z_{k-1} - f(z_{k-1})/f'(z_{k-1})$. Precision doubling motivates the following definition.

Definition. $z_0 \in \mathbb{C}$ *is an* **approximate zero** *of* f *provided*

$$\|z_k - z_{k-1}\| \le (\tfrac{1}{2})^{2^{k-1}-1}\|z_1 - z_0\|, \qquad k = 1, 2, \dots .$$

Note the ultrafast convergence due to the exponent. Toward giving a test for an approximate zero, we define a function $\alpha(z, f)$ which is central to our account.

Definition.

$$\alpha(z, f) = \left| \frac{f(z)}{f'(z)} \right| \sup_{k>1} \left| \frac{f^{(k)}(z)}{k!\, f'(z)} \right|^{1/k-1}$$

Here, $f^{(k)}(z)$ *is the kth derivative of* f *at* z.

Theorem (Special Cases).

(A) *There is a constant* α_0 *between 1/8 and 1/7 such that if* $\alpha(z, f) < \alpha_0$, *then* z *is an approximate zero of* f.

(B) $\alpha(z, f) \le \|f\|_{\max} \dfrac{|f(z)|}{|f'(z)|^2} \dfrac{\phi_d'(|z|)^2}{\phi_d(|z|)}$

Here, $\|f\|_{\max} = \sup_i |a_i|$ *and* $\phi_d(r) = \sum_{i=0}^d r^i$.

Remark. The theorem gives two tests for ensuring that Newton iteration converges fast starting immediately and continuing indefinitely. Combining (B) with (A) gives a test, while cruder than (A) alone, involving only the first derivative. These tests can be used to terminate zero finding algorithms, as well as for constructing global algorithms.

Kim (1985) has independently given a proof of Theorem (A) with $\alpha_0 = 1/54$. Since her proof uses the theory of Schlicht functions, it does not extend to several variables.

The theorem can be used to study the likelihood of z being an approximate zero for f as follows. Let P_d be the space of polynomials f of degree less than or equal to d whose coefficients satisfy $|a_i| \le 1$, $i = 0, \dots, d$. Then $P_d \subset \mathbb{C}^{d+1}$ and one can use as a probability measure on P_d normalized Lebesgue measure. Let $p(z, d)$ be the probability that z is an approximate zero of f drawn randomly from P_d.

Corollary. *That $z = 0$ is an approximate zero for f has a probability bounded below by a positive constant c independent of d. That is, $p(0, d) > c > 0$.*

Proof. Put $z = 0$ in (B) of the theorem to see that $|a_0|/|a_1|^2 < \alpha_0$ implies 0 is an approximate zero of $f(z) = \sum_0^d a_i z^i$. But $|a_0|/|a_1|^2$ is independent of d.

Remark. In fact one computes $c = \alpha_0^2/3$ by integration. The corresponding bound for real polynomials is $\alpha_0/3$.

After I found the above result I raised the question for general z. Renegar then showed using the above theorem that $p(z, d) > c(|z|) > 0$ if $|z| < 1$ and showed that no such bound existed, independent of d, with $|z| > 1$.

Section 2.

A standing hypothesis in this section is that $f : \mathcal{E} \to \mathcal{F}$ is an analytic map from one Banach space to another, both $\mathcal{E}$ and $\mathcal{F}$ are real or both are complex. Main examples are the finite dimensional cases $\mathcal{E} = \mathbb{C}^n$, $\mathcal{F} = \mathbb{C}^n$, where n could be one as in Section 1. The map f could be given by a system of polynomials.

Our calculus notation follows [5] or [1], where the basic theorems of calculus on Banach spaces which we use can also be found. Taylor series expansions of an analytic map are the most important of these.

The derivative of $f : \mathcal{E} \to \mathcal{F}$ at $z \in \mathcal{E}$ is a linear map $Df(z) : \mathcal{E} \to \mathcal{F}$. If $Df(z)$ is invertible, Newton's method provides a new vector z' from z by

$$z' = z - Df(z)^{-1}f(z) = N_f(z) = N(z).$$

Let β denote the norm of this Newton step $z' - z$, i.e.,

$$\beta(z, f) = \beta(z) = \|Df(z)^{-1}f(z)\|.$$

In case $Df(z)$ is not invertible, let $\beta(z) = \infty$.

For a point $z_0 \in \mathcal{E}$, define inductively the sequence $z_n = z_{n-1} - Df(z_{n-1})^{-1}f(z_{n-1})$ (if possible). Say that z_0 is an **approximate zero** of f if z_n is defined for all n and satisfies:

$$\|z_n - z_{n-1}\| \leq \left(\tfrac{1}{2}\right)^{2^{n-1}-1}\|z_1 - z_0\|, \qquad \text{all } n.$$

Clearly this implies that z_n is a Cauchy sequence with a limit, say, $\varsigma \in \mathcal{E}$. That $f(\varsigma) = 0$ can be seen as follows. Since $z_{n+1} - z_n = Df(z_n)^{-1}f(z_n)$,

$$\|f(z_n)\| = \|Df(z_n)(z_{n+1} - z_n)\| \leq \|Df(z_n)\|\,\|z_{n+1} - z_n\|.$$

Take the limit as $n \to \infty$, so

$$\|f(\varsigma)\| \leq \|Df(\varsigma)\| \lim \|z_{z+1} - z_n\| = 0.$$

Note that for an approximate zero, Newton's method is superconvergent starting with the first iteration. There is no large constant on the right-hand side.

The following is very simple.

Proposition 1. *If z_0 is an approximate zero and $z_n \to \varsigma$ as $n \to \infty$, then*

$$\|z_n - \varsigma\| \leq (\tfrac{1}{2})^{2^{n-1}} \|z_1 - z_0\| K,$$

where

$$K \leq 1\tfrac{3}{4}, \qquad K = 1 + \tfrac{1}{2} + \tfrac{1}{8} + \cdots$$

For the proof, sum both sides in the definition of approximate zero.

$$\|x_N - x_\ell\| \leq \sum_{n=\ell+1}^{N} \|x_n - x_{n-1}\| \leq \|x_1 - x_0\| \sum_{n=\ell+1}^{N} (\tfrac{1}{2})^{2^{n-1}-1}$$

Let $N \to \infty$ and factor $(1/2)^{2^\ell - 1}$ from the right-hand side to obtain Proposition 1.

Toward giving criteria for z to be an approximate zero define

$$\gamma(z, f) = \sup_{k>1} \left\| Df(z)^{-1} \frac{D^k f(z)}{k!} \right\|^{1/k-1}$$

or, if $Df(z)^{-1}$ or the sup does not exist, let $\gamma(z, f) = \infty$. Here $D^k f(z)$ is the kth derivative of f at z as a k-linear map. See [1] or [5]. Also, $Df(z)^{-1} D^k f(z)$ denotes the composition. The norm is the norm as a multilinear map as defined in those references. We sometimes shorten $\gamma(z, f)$ to $\gamma(z)$ or γ in situations which make the abbreviation clear.

Define $\alpha(z, f) = \beta(z, f)\gamma(z, f)$ where β is defined earlier in the section.

Theorem A. *There is a naturally defined number α_0 approximately equal to 0.130707 such that if $\alpha(z, f) < \alpha_0$, then z is an approximate zero of f.*

The issue of sharpness is discussed later.

Suppose now $f : \mathcal{E} \to \mathcal{F}$ is a map which is expressed as $f(z) = \sum_{k=0}^{d} a_k z^k$, all $z \in \mathcal{E}$, $0 < d \leq \infty$. Here $\mathcal{E}$ and $\mathcal{F}$ are Banach spaces and a_k is a bounded symmetric k-linear map from $\mathcal{E} \times \cdots \times \mathcal{E}$ (k times) to $\mathcal{F}$. Thus $a_k z^k$ is a homogeneous polynomial of degree k, so to speak. For $\mathcal{E} = \mathbb{C}^n$, this is the case in the usual sense, and in one variable a_k is the kth coefficient (real or complex) of f. Then if d is finite, f is a polynomial. Define

$$\|f\| = \sup_{k \geq 0} \|a_k\| \qquad \text{(general case)}$$

where $\|a_k\|$ is the norm of a_k as a bounded map.

Next recall

$$\phi_d(r) = \sum_{i=0}^{d} r^i, \qquad \phi_d'(r) = \frac{d}{dr}\,\phi_d(r)$$

and $\phi(r) = \phi_\infty(r)$.

Theorem B.

$$\gamma(z) < \|Df(z)^{-1}\| \, \|f\| \, \frac{\phi_d'(\|z\|)^2}{\phi_d(\|z\|)}$$

Here, if $Df(z)$ is not invertible interpret $\|Df(z)^{-1}\| = \infty$ as usual.

Thus combining Theorems A and B, we have a first derivative criterion (at z) for z to be an approximate zero.

Corollary. *If*

$$\|f\| \, \|Df(z)^{-1}\| \, \frac{\phi_d'(\|z\|)^2}{\phi_d(\|z\|)} \, \beta(z, f) < \alpha_0$$

then z is an approximate zero of f.

Note that the norm on f treats equally all the a_k in spite of the fact that the number of monomials in a single a_k grows exponentially with k. This seems reasonable.

The proof of Theorem B will be given in a subsequent paper.

There is a reason to consider an alternate definition of approximate zero. This second notion is in terms of an actual zero ς of $f : \mathcal{E} \to \mathcal{F}$. We say that z_0 is an **approximate zero of the second kind** of $f : \mathcal{E} \to \mathcal{F}$ provided there is some $\gamma \in \mathcal{E}$ with $f(\varsigma) = 0$ and

$$\|z_n - \varsigma\| \leq (\tfrac{1}{2})^{2^{n-1}} \|z_0 - \varsigma\| \qquad \text{for } n = 1, 2, 3, \ldots$$
$$z_n = z_{n-1} - Df(z_{n-1})^{-1} f(z_{n-1})$$

While the first definition of approximate zero deals with information at hand, and computable quantities (in principle, eventually), an approximate zero of the second kind can often be studied statistically or theoretically more handily.

Theorem C. *Suppose that $f : \mathcal{E} \to \mathcal{F}$ is analytic, $\varsigma \in \mathcal{E}$, $f(\varsigma) = 0$ and $z \in \mathcal{E}$ satisfies*

$$\|z - \varsigma\| < \left(\frac{3 - \sqrt{7}}{2} \right) \frac{1}{\gamma(\varsigma)}.$$

Then z is an approximate zero of the second kind.

This result gives more evidence for the importance of the invariant $\gamma(\varsigma)$.

Section 3.

Here we prove some lemmas and propositions from which the main results will follow easily. Suppose $\mathcal{E}$ and $\mathcal{F}$ are Banach spaces, both real or both complex.

Lemma 1. *Let $A, B : \mathcal{E} \to \mathcal{F}$ be bounded linear maps with A invertible such that $\|A^{-1}B - I\| < c < 1$. Then B is invertible and $\|B^{-1}A\| < 1/(1 - c)$.*

Proof. Compare [1, 8.3.2.1]. Let $v = I - A^{-1}B$. Since $\|v\| < 1$, $\sum_0^\infty v^i$ exists with norm $< 1/(1 - c)$. Note $(I - v) \sum_0^n v^i (I - v) = I - v^{n+1}$. By taking limits, $A^{-1}B = I - v$ is seen to be invertible with inverse $\sum_0^\infty v^i$. Since $A^{-1}B \sum_0^\infty v^i = I$, B can be written as the composition of invertible maps. ∎

Lemma 2. *Suppose $f : \mathcal{E} \to \mathcal{F}$ is analytic, $z', z \in \mathcal{E}$ such that $\|z' - z\|\gamma(z) < 1 - \sqrt{2}/2$. Then*

(a) $Df(z')$ is invertible.

(b) $\|Df(z')^{-1}Df(z)\| < \dfrac{1}{2 - \phi'(\|z' - z\|\gamma(z))}$

(c) $\gamma(z') \leq \gamma(z) \dfrac{1}{2 - \phi'(\|z' - z\|\gamma(z))} \left(\dfrac{1}{1 - \|z' - z\|\gamma(z)}\right)^3$

Here $\phi'(r) = 1/(1 - r)^2$ could be replaced by ϕ'_d.

Proof. Take a Taylor series expansion of Df about z (i.e., the map $\mathcal{E} \to L(\mathcal{E}, \mathcal{F})$) as follows (see [1], [5]).

$$Df(z') = \sum_{k=0}^{\infty} \frac{D^{k+1}f(z)}{k!} (z' - z)^k$$

From this,

$$Df(z)^{-1}Df(z') = \sum_{k=0}^{\infty} Df(z)^{-1} \frac{D^{k+1}f(z)}{k!} (z' - z)^k$$

$$Df(z)^{-1}Df(z') - I = \sum_{k=1}^{\infty} (k + 1) \frac{Df(z)^{-1} D^{k+1}f(z)}{(k + 1)!} (z' - z)^k$$

and

$$\|Df(z)^{-1}Df(z') - I\| \leq \sum_{k=1}^{\infty} (k + 1) \left\| \frac{Df(z)^{-1} D^{k+1}f(z)}{(k + 1)!} \right\| \|z' - z\|^k$$

$$\leq \sum_{k=1}^{\infty} (k + 1) \left(\gamma(z)\|z' - z\|\right)^k$$

$$\leq \phi'\left(\gamma(z)\|z' - z\|\right) - 1.$$

Observe, that since $\gamma(z)\|z' - z\| < 1 - \sqrt{2}/2$ all the series converge. Moreover, note that $\phi'(r) = 1/(1 - r)^2$, so that if $r < 1 - \sqrt{2}/2$, $\phi'(r) - 1 < 1$. Thus Lemma 1 applies to yield parts (a) and (b) of Lemma 2.

The following simple formulae come up frequently and we use them without comment.

$$\frac{\phi^{(\ell)}}{\ell!}(r) = \sum_{k=0}^{\infty} \binom{\ell + k}{k} r^k = \frac{1}{(1 + r)^{\ell+1}}.$$

Both quantities on the right are seen to be equal to the ℓth derivative.

$$\left(\frac{1}{2 - \phi'(r)}\right)\left(\frac{1}{(1 - r)^2}\right) = \frac{1}{\psi(r)}$$

where

$$\psi(r) = 2r^2 - 4r + 1.$$

Now we prove (c) of Lemma 2. Let

$$\gamma_k = \gamma_k(z) = \left\| Df(z)^{-1} \frac{D^{(k)}f(z)}{k!} \right\|^{1/k-1} \qquad \text{and} \qquad \gamma = \sup_{k>1} \gamma_k.$$

Then by Taylor's theorem (see [5])

$$\gamma_k(z')^{k-1} = \left\| Df(z')^{-1} Df(z) \sum_{\ell=0} \frac{Df(z)^{-1} D^{k+\ell}f(z)(z'-z)^\ell}{\ell!\,k!} \right\|$$

$$\le \|Df(z')^{-1}Df(z)\| \sum_{\ell=0} \binom{k+\ell}{\ell} \left\| \frac{Df(z)^{-1}D^k f(z)(z'-z)^\ell}{(k+\ell)!} \right\|$$

$$\le \|Df(z')^{-1}Df(z)\| \gamma(z)^{k-1} \sum_{\ell=0} \binom{k+\ell}{\ell} \left(\|z'-z\|\gamma(z) \right)^\ell$$

$$\le \|Df(z')^{-1}Df(z)\| \gamma^{k-1} \left(\frac{1}{1-\gamma\|z'-z\|} \right)^{k+1}$$

Now use Lemma 2b and take the $(k-1)$ root to obtain

$$\gamma_k(z') \le \frac{\gamma}{2 - \phi'(\gamma\|z'-z\|)} \left(\frac{1}{1-\gamma\|z'-z\|} \right)^{(k+1)/(k-1)}.$$

The supremum is achieved at $k = 2$, yielding the statement of Lemma 2c.

Lemma 3.

(a) Let $\alpha = \alpha(z,f) < 1$ and $z' = z - Df(z)^{-1}f(z)$, $\beta = \beta(z,f)$. Then

$$\|Df(z)^{-1}f(z')\| \le \beta\left(\frac{\alpha}{1-\alpha} \right).$$

(b) Let $z, z' \in \mathcal{E}$ with $f(z) = 0$, and $\|z'-z\|\gamma(z) < 1$. Then

$$\|Df(z)^{-1}f(z')\| \le \frac{\|z'-z\|}{1 - \|z'-z\|\gamma(z)}$$

We first prove part (a).

The Taylor series yields

$$Df(z)^{-1}f(z') = \sum_{k=0} Df(z)^{-1} \frac{D^k f(z)(z'-z)^k}{k!}.$$

The first two terms drop out. Since $\beta(z) = \|z'-z\|$,

$$\|Df(z)^{-1}f(z')\| \le \beta \sum_{k=2} (\gamma\beta)^{k-1} \le \beta \frac{\alpha}{1-\alpha}$$

proving (a).

For part (b), we start as above and now the first term is zero since $f(z) = 0$. We have

$$\|Df(z)^{-1}f(z')\| \le \|(z'-z)\| \left(1 + \sum_{k=2} \gamma_k^{k-1} \|z'-z\|^{k-1} \right)$$

$$\le \|z'-z\| \sum_{k=0} (\gamma\|z'-z\|)^k$$

$$\le \|z'-z\| \frac{1}{1-\gamma\|z'-z\|}.$$

This finishes the proof of Lemma 3.

Proposition 1. *Let* $\alpha = \alpha(z) = \alpha(z, f)$, $\beta = \beta(z)$, $\beta' = \beta(z')$ *where* $f : \mathcal{E} \to \mathcal{F}$ *is an analytic map from the Banach space* $\mathcal{E}$ *to* $\mathcal{F}$ *as usual.*

(a) if $\alpha < 1 - \sqrt{2}/2$, *then*

$$\beta' \leq \beta \left(\frac{\alpha}{1 - \alpha} \right) \left(\frac{1}{2 - \phi'(\alpha)} \right)$$

(b) if $f(z) = 0$ *and* $\gamma \|z' - z\| < 1 - \sqrt{2}/2$ *then*

$$\beta' \leq \|z' - z\| \frac{1}{2 - \phi'(\gamma \|z' - z\|)} \left(\frac{1}{1 - \gamma \|z' - z\|} \right)$$

Proof of Proposition 1. Write

$$\beta(z') = \|Df(z')^{-1} f(z')\| = \|Df(z')^{-1} Df(z) Df(z)^{-1} f(z')\|$$
$$\leq \|Df(z')^{-1} Df(z)\| \, \|Df(z)^{-1} f(z')\|$$
$$\leq \left(\frac{1}{2 - \phi'(\alpha)} \right) \beta \left(\frac{\alpha}{1 - \alpha} \right).$$

The last step uses Lemmas 2b and 3a.

Similarly for the second part of the proposition. If $f(z) = 0$, use Lemmas 2b and 3b as follows.

$$\beta(z') \leq \|Df(z')^{-1} Df(z)\| \, \|Df(z)^{-1} f(z')\|$$
$$\leq \frac{1}{2 - \phi'(\|z' - z\| \gamma(z))} \|z' - z\| \frac{1}{\left(1 - \|z' - z\| \gamma(z) \right)}$$

This proves Proposition 1.

Proposition 2. *Recalling* $\psi(r) = 2r^2 - 4r + 1$ *and using the notation of Proposition 1,*

(a) if $\alpha < 1 - \sqrt{2}/2$,

$$\alpha' \leq \left(\frac{\alpha}{\psi(\alpha)} \right)^2$$

(b) if $f(\varsigma) = 0$ *and* $\|z - \varsigma\| \gamma(\varsigma) < 1 - \sqrt{2}/2$,

$$\alpha' \leq \frac{\gamma(\varsigma) \|z - \varsigma\|}{\psi \left(\gamma(\varsigma) \|z - \varsigma\| \right)^2}$$

For the proof of (a), note that $\alpha' = \beta' \gamma'$. Use Lemma 2c and Proposition 1a to obtain

$$\alpha' \leq \beta \left(\frac{\alpha}{1 - \alpha} \right) \left(\frac{1}{2 - \phi'(\alpha)} \right) \gamma \left(\frac{1}{1 - \alpha} \right)^3 \left(\frac{1}{2 - \phi'(\alpha)} \right)$$
$$\alpha' \leq (\beta \gamma) \alpha \left(\frac{1}{\psi(\alpha)} \right)^2 \leq \left(\frac{\alpha}{\psi(\alpha)} \right)^2$$

The proof of Proposition 2b is similar using Lemma 2c and Proposition 1b.

The following is proved very simply by induction.

Proposition 3. *Suppose that $A > 0$, $a_i > 0$, $i = 0, 1, \ldots$ satisfy: $a_{i+1} \leq A a_i^2$, all i.*
Then

$$a_k \leq (A a_0)^{2^k - 1} a_0, \qquad \text{all } k.$$

We end this section with a short discussion of sharpness. Lemma 2b can be seen to be sharp by taking $z = 0$,

$$f(z) = 2z - \phi(z) + 1 - a, \qquad 0 < a < 1 - \frac{\sqrt{2}}{2}.$$

Then $z' = a$, $\gamma(0, f) = 1$ and $|z' - z\gamma(0, f)| = a$ at $z = 0$.

The same example may be used to see that Lemma 2c is sharp. One only needs to make the easy computation that

$$\gamma(a) = \left(\frac{1}{1 - \alpha}\right)^3 \left(\frac{1}{2 - \phi'(a)}\right)$$

Again, the same example shows that Lemma 3a is sharp, just observing that $\alpha(0, f) = a$, $f(a) = a^2/(1 - a)$.

One can see that Lemma 3b is sharp with the example

$$f(z) = \phi(z) - 1, \qquad z = 0, \qquad 0 < z' < 1.$$

Proposition 1a is sharp with the example of Lemma 2. The same applies to Proposition 2a.

Section 4.

In this final section we finish the proofs of our main results.

Toward the proof of Theorem A, consider our polynomial $\psi(r) = 2r^2 - 4r + 1$, and the function $\left(\alpha/\psi(\alpha)\right)^2$ of Proposition 2a. In the range of concern to us, $0 \leq r \leq 1 - \sqrt{2}/2$, $\psi(r)$ is a parabola decreasing from 1 to 0 as r goes from 0 to $1 - \sqrt{2}/2$. Therefore $r/\psi(r)^2$ increases from 0 to ∞ as r goes from 0 to $1 - \sqrt{2}/2$. Let α_0 be the unique r such that $r/\psi(r)^2 = 1/2$. Thus α_0 is a zero of the real quartic polynomial $\psi(r)^2 - 2r$. Using Newton's method one calculates approximately $\alpha_0 = 0.130707$. With this discussion, Theorem A is a consequence of the following proposition where $a = 1/2$.

Proposition 1. *Let $f : \mathcal{E} \to \mathcal{F}$ be analytic, $z = z_0 \in \mathcal{E}$ and $\alpha(z)/\psi(\alpha(z))^2 = a < 1$. Let $z_k = z_{k-1} - D(z_{k-1})^{-1} f(z_{k-1})$, $k = 1, 2, \ldots$, then*
(a) z_k is defined for all k.
Let $\alpha_k = \alpha(z_k)$, $\psi_k = \psi(\alpha(z_k))$, $k = 1, 2, \ldots$,
(b) $\alpha_k \leq a^{2^k - 1} \alpha(z_0)$, $k = 1, 2, \ldots$,
(c) $\|z_k - z_{k-1}\| \leq a^{2^{k-1} - 1} \|z_1 - z_0\|$, all k.

Proof. Note that (a) follows from (b) and that (b) is a consequence of Propositions 2a and 3 of Section 3. It remains to check (c). The case $k = 1$ is trivial so assume $k > 1$.

We may write using Proposition 1a and our relation between ϕ' and ψ,

$$\|z_n - z_{n-1}\| \leq \|z_{n-1} - z_{n-2}\| \frac{\alpha_{n-2}(1 - \alpha_{n-2})}{\psi_{n-2}}$$

Now use part (b) and induction on this inequality to obtain

$$\|z_n - z_{n-1}\| \leq a^{2^{n-2}-1}\|z_1 - z_0\| a^{2^{n-1}-1} \alpha(z_0)\left(\frac{1 - \alpha_{n-2}}{\psi_{n-2}}\right)$$

$$\leq a^{2^{n-1}-1}\|z_1 - z_0\| \frac{\alpha(z_0)}{\psi_{n-2}}$$

But

$$\frac{\alpha(z_0)}{\psi_{n-2}} \leq \frac{\alpha_0}{\psi_0} \leq a\psi_0 < 1.$$

Remark. Theorem A and most of the lemmas and propositions of Section 3 can be slightly sharpened in case that f is a polynomial map $\mathcal{E} \to \mathcal{F}$ of Banach spaces of degree $d < \infty$. Replace $\phi(r)$ by $\phi_d(r) = \sum_0^d r^i$ everywhere in the proofs and conclusions.

For example, going through proofs this way yields the following generalization of Proposition 2a of Section 3.

Proposition 2. *If $f : \mathcal{E} \to \mathcal{F}$ has degree d, then (using notations of Section 3) if $\phi_d'(\alpha) < 2$,*

$$\alpha' < \alpha^2 \frac{\phi_{d-2}(\alpha)}{\phi_d(\alpha)}\left(\frac{\phi_d'(\alpha)}{2 - \phi_d'(\alpha)}\right)^2$$

If $d = \infty$, this reverts to Proposition 2a, Section 3.

Example. The following shows that α_0 must be less than or equal $3 - 2\sqrt{2}$ in Theorem A. Let $f_a : \mathbb{C} \to \mathbb{C}$ be $f_a(z) = 2z - z/(1 - z) - a$, $a > 0$. Then $\alpha(0, f_a) = a$ and $f_a(\varsigma) = 0$ where $\varsigma = \left((1 + a) \pm \sqrt{(1 + a)^2 - 8a}\right)/4$. If $\alpha = a > 3 - 2\sqrt{2}$, these roots are not real, so that Newton's method for solving $f_a(\varsigma) = 0$, starting at $z_0 = 0$ will never converge.

Toward the proof of Theorem C, we have:

Proposition 3. *Let $f : \mathcal{E} \to \mathcal{F}$, $\varsigma, z \in \mathcal{E}$ satisfy $f(\varsigma) = 0$ and $\gamma(\varsigma)\|z - \varsigma\| < 1 - \sqrt{2}/2$. Then*

$$\|Df(z)^{-1}f(z) - (z - \varsigma)\| \leq \frac{\gamma(\varsigma)\|z - \varsigma\|^2}{\psi(\gamma(\varsigma)\|z - \varsigma\|)}$$

Note that this proposition gives an estimate on how well the Newton vector $-Df(z)^{-1}f(z)$ approximates $\varsigma - z$, the exact vector from z to ς.

For the proof we consider the two Taylor series:

$$f(z) = \sum_{k=1} \frac{D^k f(\varsigma)}{k!}(z - \varsigma)^k$$

and

$$Df(z) = \sum_{k=0} \frac{D^{k+1}f(\varsigma)}{k!} (z - \varsigma)^k$$

Now apply the second to $(z - \varsigma)$ and subtract it from the first to obtain:

$$f(z) - Df(z)(z - \varsigma) = -\sum_{k=1}(k - 1)\frac{D^k f(\varsigma)}{k!} (z - \varsigma)^k$$

Then

$$Df(z)^{-1}f(z) - (z - \varsigma) = -Df(z)^{-1}Df(\varsigma) \sum_{1}(k - 1)\frac{Df(\varsigma)^{-1}D^k f(\varsigma)(z - \varsigma)^k}{k!}$$

Take norms and apply Lemma 2b to obtain:

$$\|Df(z)^{-1}f(z) - (z - \varsigma)\| \leq \left(\frac{1}{2 - \phi'(w)}\right)\left(\sum_{2}(k - 1)w^{k-1}\right)\|z - \varsigma\|$$

where $w = \gamma(\varsigma)\|z - \varsigma\|$.

The right hand is estimated by $\|z - \varsigma\|w/\psi(w)$ proving the proposition.

Corollary. *Suppose f, ς, z are as in the proposition. Let*

$$A = \frac{\gamma(\varsigma)\|z - \varsigma\|}{\psi(\gamma(\varsigma)\|z - \varsigma\|)} < 1$$

or equivalently

$$\|z - \varsigma\| < \frac{5 - \sqrt{17}}{4}\left(\frac{1}{\gamma(\varsigma)}\right).$$

Then

$$\|z_n - \varsigma\| \leq A^{2^n - 1}\|z - \varsigma\|$$

where $z = z_0$, $z_n = z_{n-1} - Df(z_{n-1})^{-1}f(z_{n-1})$.

This follows from Proposition 3 using Proposition 3 of the previous section.

Now Theorem C follows by choosing $A = 1/2$ in the corollary.

Traub and Wozniakowski (1979) may be reinterpreted as lying in the direction of this corollary.

For the sharpness of the corollary, consider $f(z) = z/(1 - z)$ with $\varsigma = 0$. Then $\gamma(\varsigma) = 1$ and $z_n = z_{n-1}^2$.

Acknowledgment. This work was supported in part by NSF and DOE contracts.

References.

[1] J. Dieudonné, *Foundations of Modern Analysis*, Academic Press, New York, 1960.

[2] W. Gregg and R. Tapia, "Optimal error bounds for the Newton-Kantorovich theorem," *SIAM J. Numer. Anal.*, **11**(1974), 10–13.

[3] L. Kantorovich and G. Akilov, *Functional Analysis in Normed Spaces*, MacMillan, New York, 1964.

[4] M. Kim, Ph.D. Thesis, City University of New York, to appear, 1985.

[5] S. Lang, *Real Analysis*, Addison-Wesley, Reading, Massachusetts, 1983.

[6] A. Ostrowski, *Solutions of Equations in Euclidean and Banach Spaces*, Academic Press, New York, 1973.

[7] L. Rall, "A note on the convergence of Newton's method," *SIAM J. Numer. Anal.*, **11**(1974), 34–36.

[8] J. Renegar, "On the efficiency of Newton's method in approximating all zeroes of a system of complex polynomials," *Mathematics of Operations Research*, to appear.

[9] M. Shub and S. Smale, "Computational complexity: on the geometry of polynomials and a theory of cost, part I," *Ann. Sci. Ecole Norm. Sup. 4 serie t*, **18**(1985), 107–142.

[10] M. Shub and S. Smale, "Computational complexity: on the geometry of polynomials and a theory of cost, part II," *SIAM J. Computing*, **15**(1986), 145–161.

[11] S. Smale, "The fundamental theorem of algebra and complexity theory," *Bull. Amer. Math. Soc. (N.S.)*, **4**(1981), 1–36.

[12] S. Smale, "On the efficiency of algorithms of analysis," *Bull. Amer. Math. Soc. (N.S.)*, **13**(1985), 87–121.

[13] J. Traub and H. Wozniakowski, "Convergence and complexity of Newton iteration for operator equations," *J. Assoc. Comp. Mach.*, **29**(1979), 250–258.

The author has been at Berkeley since 1964. His work is in the analysis of algorithms. He spends his spare time in the pursuit of collecting fine mineral specimens, and sailing in San Francisco Bay and the adjoining ocean.

ERROR BOUNDS FOR NEWTON'S METHOD
UNDER THE KANTOROVICH ASSUMPTIONS

TETSURO YAMAMOTO

Department of Mathematics
Faculty of Science
Ehime University
Matsuyama 790, Japan

ABSTRACT

A principle which finds sharp a posteriori error bounds for Newton's method is given under the assumptions of Kantorovich's theorem. On the basis of this principle, new error bounds are derived and comparison is made among the known bounds of Dennis [1], Gragg-Tapia [4], Kantorovich [5], [6], Kornstaedt [8], Lancaster [9], Miel [11], Moret [13], Ostrowski [16], [17], Potra [18], and Potra-Pták [19].

AMS(MOS) Subject Classifications: 65G99, 65J15

Key Words: Error estimates, Newton's method, Kantorovich's theorem, Kantorovich's assumptions

1. Introduction.

The Newton method is one of the central numerical techniques for solving nonlinear equations in Euclidean or Banach spaces. Basic results concerning convergence of the method, error estimates and existence and uniqueness regions of solutions were given by Kantorovich [5], [6], (also see [7], [15], [21]). Since then, a large number of results [1]–[4], [8]–[12], [14]–[20], [22], [23], [26] have been published concerning convergence and error estimates for the Newton method under the assumptions of the Kantorovich theorem or under closely related assumptions. Recently, in a series of papers [27]–[30], we compared the upper and lower bounds obtained by Dennis [1], Gragg-Tapia [4], Miel [11], [12], Moret [13], Potra-Pták [19] and Schmidt [25].

In this paper, we present a new principle which finds sharp a posteriori error bounds for the Newton method under the Kantorovich assumptions. On the basis of this principle, new error bounds are derived, which improve their bounds as well as those obtained by Kornstaedt [8], Lancaster [9] and recently by Potra [18]. Our approach certainly simplifies and unifies the study of finding sharp error bounds for the Newton method under the Kantorovich assumptions.

2. Preliminaries.

Let X and Y be Banach spaces and $F : D \subseteq X \to Y$ be Fréchet differentiable in an open convex set $D_0 \subseteq D$. Assume that for some $x_0 \in D_0$, $F'(x_0)^{-1} \in L(Y, X)$ exists, where $L(Y, X)$ denotes the Banach space of bounded linear operators of Y into X. Then, the Newton method for solving the equation

$$F(x) = 0, \tag{2.1}$$

which starts from the initial approximation x_0, is defined by

$$x_{n+1} = x_n - F'(x_n)^{-1} F(x_n), \qquad n \geq 0, \tag{2.2}$$

provided that $F'(x_n)^{-1} \in L(Y, X)$ exists at each step. Then, convergence of the method, error estimates and existence and uniqueness regions of solutions are given by the following Kantorovich theorem, which we state under the slightly weaker assumptions than the original one (cf. Schmidt [24], Yamamoto [28; Remark 2] and [31; p. 13]).

Theorem 2.1 (Kantorovich [5], [6], Kantorovich-Akilov [7]). *Let $F : D \subseteq X \to Y$ be Fréchet differentiable in an open convex set $D_0 \subseteq D$, and, for some $x_0 \in D_0$, $F'(x_0)^{-1}$ exist. Assume that $F(x_0) \neq 0$ without loss of generality and that*

$$\left\| F'(x_0)^{-1} \big(F'(x) - F'(y) \big) \right\| \leq K \| x - y \|, \qquad x, y \in D_0, \quad K > 0,$$

$$\eta = \| F'(x_0)^{-1} F(x_0) \|, \qquad h = K\eta \leq \tfrac{1}{2}, \qquad t^* = \frac{2\eta}{1 + \sqrt{1 - 2h}},$$

$$\bar{S} = \bar{S}(x_1, t^* - \eta) = \{ x \in X \mid \| x - x_1 \| \leq t^* - \eta \} \subseteq D_0. \tag{2.3}$$

Then:

(i) The iterates (2.2) are well-defined, lie in $\bar{S}$ and converge to a solution x^* of (2.1).

(ii) The solution is unique in $S(x_0, t^{**}) \cap D_0$, $t^{**} = (1 + \sqrt{1 - 2h})/K$ if $2h < 1$, and in $\bar{S}(x_0, t^{**}) \cap D_0$ if $2h = 1$, where $S(x_0, t^{**}) = \{ x \in X \mid \| x - x_0 \| < t^{**} \}$ and $\bar{S}(x_0, t^{**})$ stands for its closure.

(iii) Error estimates

$$\| x^* - x_n \| \leq \frac{2\eta_n}{1 + \sqrt{1 - 2h_n}} \leq 2^{1-n} (2h)^{2^n - 1} \eta, \qquad n \geq 0, \tag{2.4}$$

and

$$\|x^* - x_n\| \le t^* - t_n, \qquad n \ge 0 \tag{2.5}$$

hold, where $\{\eta_n\}$ and $\{h_n\}$ are defined by the recurrence relations

$$B_0 = 1, \qquad \eta_0 = \eta, \qquad h_0 = h = K\eta,$$

$$B_n = \frac{B_{n-1}}{1 - h_{n-1}}, \qquad \eta_n = \frac{h_{n-1}\eta_{n-1}}{2(1 - h_{n-1})}, \qquad h_n = KB_n\eta_n, \qquad n \ge 1,$$

and $\{t_n\}$ is defined by

$$t_0 = 0, \qquad t_{n+1} = t_n - \frac{f(t_n)}{f'(t_n)}, \qquad n \ge 0$$

with $f(t) = \frac{1}{2}Kt^2 - t + \eta$.

Proof. The proof is done along the same line as in Kantorovich [5] and [6]. In fact, it is easy to see that, by the standard majorant theory due to Kantorovich, we have

$$\|x_{n+1} - x_n\| \le t_{n+1} - t_n, \qquad n \ge 0.$$

Also see the proof of [31; Theorem 3.1]. ∎

Kantorovich established the bounds (2.4) in his paper [5] and, three years later, obtained the bound (2.5) in [6]. However, it can be shown [29], [30] that

$$\frac{2\eta_n}{1 + \sqrt{1 - 2h_n}} = t^* - t_n, \qquad n \ge 0.$$

With this remarkable property, we state the following lemmas, whose proofs are essentially the same as in [30].

Lemma 2.1. *Under the assumptions of Theorem 2.1, put* $\Delta = t^{**} - t^*$ *and* $\theta = t^*/t^{**}$. *Then*

$$t_{n+1} - t_n = \eta_n = \begin{cases} \dfrac{\Delta\theta^{2^n}}{1 - \theta^{2^{n+1}}} & (2h < 1) \\ 2^{-n}\eta & (2h = 1) \end{cases}, \tag{2.6}$$

$$t^* - t_n = \frac{2\eta_n}{1 + \sqrt{1 - 2h_n}} = \begin{cases} \dfrac{\Delta\theta^{2^n}}{1 - \theta^{2^n}} & (2h < 1) \\ 2^{1-n}\eta & (2h = 1) \end{cases}, \tag{2.7}$$

$$t^{**} - t_n = \frac{1 + \sqrt{1 - 2h_n}}{K} = \frac{\Delta}{1 - \theta^{2^n}}, \qquad (2h < 1).$$

That is, $t^* - t_n$ *and* $t^{**} - t_n$ *are the solutions of the equation* $\frac{1}{2}KB_nt^2 - t + \eta_n = 0$.

Lemma 2.2. *Let the assumptions of Theorem 2.1 hold. Then:*

(i) $\quad B_n = (1 - Kt_n)^{-1} = \sqrt{1 - 2h + (K\eta_{n-1})^2}^{-1}$

$$= \left\{ K\eta_n + \sqrt{1 - 2h + (K\eta_n)^2} \right\}^{-1}$$

$$= \begin{cases} \dfrac{\sqrt{1 - 2h_n}}{\sqrt{1 - 2h}} = \dfrac{2}{\Delta K} \cdot \dfrac{1 - \theta^{2^n}}{1 + \theta^{2^n}} & (2h < 1) \\ 2^n & (2h = 1) \end{cases}, \quad (n \ge 0)$$

(ii)
$$\frac{t^* - t_{n+1}}{(t^* - t_n)^2} = \frac{\nabla t_{n+1}}{(\nabla t_n)^2} = \frac{1}{2} K B_n \qquad (n \geq 1),$$

(iii)
$$\frac{t^* - t_n}{(\nabla t_n)^2} = \frac{K B_n}{1 + \sqrt{1 - 2h_n}} = \begin{cases} \dfrac{1 - \theta^{2^n}}{\Delta} & (2h < 1) \\ \dfrac{2^{n-1}}{\eta} & (2h = 1) \end{cases} \quad (n \geq 1),$$

where $\nabla t_n = t_n - t_{n-1}$.

Lemma 2.3. *Under the assumptions of Theorem 2.1, choose $\phi \geq 0$ such that $(1 + \cosh\phi)h = 1$. Then:*

(i) $\quad t^* - t_n = \begin{cases} e^{-2^{n-1}\phi} \dfrac{\sinh\phi}{\sinh 2^{n-1}\phi} \eta & (2h < 1) \\ 2^{1-n}\eta = \lim\limits_{\phi \to +0} \left(e^{-2^{n-1}\phi} \dfrac{\sinh\phi}{\sinh 2^{n-1}\phi} \eta \right) & (2h = 1) \end{cases} \quad (n \geq 0),$

(ii)
$$\frac{t^* - t_n}{\nabla t_n} = \theta^{2^{n-1}} = e^{-2^{n-1}\phi} \qquad (n \geq 1).$$

Lemma 2.4. *Under the assumptions of Theorem 2.1, we have*

$$t^* - t_n = \frac{1 - \sqrt{1 - 2h_n}}{K B_n} \leq \frac{1}{2^n K} (1 - \sqrt{1 - 2h})^{2^n}, \qquad n \geq 0.$$

Lemma 2.5. *Under the assumptions of Theorem 2.1, put $d_n = \|x_{n+1} - x_n\|$ and $a = \sqrt{1 - 2h}/K$. Then, for $n \geq 1$, we have:*

(i)
$$d_n \leq \frac{d_{n-1}^2}{2\sqrt{a^2 + \eta_{n-1}^2}} \leq \frac{d_{n-1}^2}{2\sqrt{a^2 + d_{n-1}^2}} \leq \frac{1}{2} d_{n-1},$$

(ii)
$$d_n + \sqrt{a^2 + d_n^2} \leq \sqrt{a^2 + d_{n-1}^2},$$

(iii)
$$\frac{d_{n-1}}{2\sqrt{a^2 + d_{n-1}^2}} \leq \frac{\eta_n}{\eta_{n-1}} = \frac{\nabla t_{n+1}}{\nabla t_n}.$$

3. Error Bounds for $\|x^* - x_n\|$.

We can prove the following results, whose proofs will be given in [32].

Theorem 3.1 (A Principle To Find A Posteriori Error Bounds). *Let the assumptions of Theorem 2.1 hold. Furthermore, assume that for some $n \geq 0$, there exist nonnegative numbers M_n and N_n such that $M_n \leq N_n \leq KB_n$ and*

$$\|x^* - x_{n+1}\| \leq \tfrac{1}{2}M_n\|x^* - x_n\|^2.$$

Then we have

$$\tau_n^* \equiv \frac{2d_n}{1 + \sqrt{1 + 2N_n d_n}} \leq \sigma_n^* \equiv \frac{2d_n}{1 + \sqrt{1 + 2M_n d_n}}$$

$$\leq \quad \|x^* - x_n\| \quad \leq \bar\sigma_n^* \equiv \frac{2d_n}{1 + \sqrt{1 - 2M_n d_n}}$$

$$\leq \bar\tau_n^* \equiv \frac{2d_n}{1 + \sqrt{1 - 2N_n d_n}}$$

$$\leq \frac{t^* - t_n}{\nabla t_{n+1}}d_n \leq t^* - t_n, \qquad n \geq 0,$$

where $d_n = \|x_{n+1} - x_n\|$.

Corollary 3.1.1. *With the notation and assumptions of Theorem 2.1, we have*

$$\underline{\tau}_n \equiv \frac{2d_n}{1 + \sqrt{1 + 2KB_n}} \leq \underline{\sigma}_n \equiv \frac{2d_n}{1 + \sqrt{1 + 2K(1 - K\Delta_n)^{-1}d_n}}$$

$$\leq \quad \|x^* - x_n\| \quad \leq \sigma_n \equiv \frac{2d_n}{1 + \sqrt{1 - 2K(1 - K\Delta_n)^{-1}d_n}}$$

$$\leq \tau_n \equiv \frac{2d_n}{1 + \sqrt{1 - 2KB_n d_n}}, \qquad n \geq 0,$$

where $\Delta_n = \|x_n - x_0\|$.

Theorem 3.2. *Under the assumptions of Theorem 2.1, put*

$$\bar S_0 = \bar S, \qquad \bar S_n = \bar S(x_n, t^* - t_n) \quad (n \geq 1), \qquad K_0 = L_0 = K,$$

$$K_n = \sup_{\substack{x,y\in\bar S_n \\ x\neq y}} \frac{\|F'(x_n)\big(F'(x) - F'(y)\big)\|}{\|x - y\|} \qquad (n \geq 1),$$

$$L_n = \sup_{\substack{x,y\in\bar S \\ x\neq y}} \frac{\|F'(x_n)^{-1}\big(F'(x) - F'(y)\big)\|}{\|x - y\|} \qquad (n \geq 1). \tag{3.1}$$

Then

$$\underline{\lambda}_n \equiv \frac{2d_n}{1 + \sqrt{1 + 2L_n d_n}} \leq \underline{\kappa}_n \equiv \frac{2d_n}{1 + \sqrt{1 + 2K_n d_n}} \tag{3.2}$$

$$\leq \quad \|x^* - x_n\| \quad \leq \kappa_n \equiv \frac{2d_n}{1 + \sqrt{1 - 2K_n d_n}} \tag{3.3}$$

$$\leq \lambda_n \equiv \frac{2d_n}{1 + \sqrt{1 - 2L_n d_n}}$$

$$\leq \frac{t^* - t_n}{\nabla t_{n+1}} d_n \leq t^* - t_n, \qquad n \geq 0.$$

Remark 3.1. The constants L_n were first defined by Lancaster [9] and later by Kornstaedt [8], Ypma [33], etc.

Remark 3.2. If we replace the condition (2.3) in Theorem 2.1 by the stronger one $\bar{S}' = \bar{S}(x_1, \eta) \subseteq D_0$, and put

$$\bar{S}'_0 = \bar{S}', \qquad \bar{S}'_n = \bar{S}(x_n, 2d_n) \qquad (n \geq 1),$$

$$K'_n = \sup_{\substack{x, y \in \bar{S}'_n \\ x \neq y}} \frac{\left\| F'(x_n)^{-1}\left(F'(x) - F'(y)\right) \right\|}{\|x - y\|} \qquad (n \geq 0),$$

then we have

$$x^* \in \bar{S}'_n \subseteq \bar{S}'_{n-1} \subseteq \cdots \subseteq \bar{S}'_0$$

and

$$\underline{\kappa}'_n \equiv \frac{2d_n}{1 + \sqrt{1 + 2K'_n d_n}} \leq \|x^* - x_n\| \leq \kappa'_n \equiv \frac{2d_n}{1 + \sqrt{1 - 2K'_n d_n}} \qquad (n \geq 0).$$

Theorem 3.1 or 3.2 gives us a unified technique for finding error bounds for Newton's method under the Kantorovich assumptions. In fact, we have

$$K_n \leq L_n \leq K(1 - K\Delta_n)^{-1} \leq K(1 - Kt_n)^{-1} = KB_n.$$

Hence, from Theorem 3.2, Corollary 3.1.1 and Lemmas 2.1–2.5, we have the following chart for the upper bounds for the error $\|x^* - x_n\|$:

$$\|x^* - x_n\| \leq \kappa_n \quad (n \geq 0) \qquad \leq \lambda_n \quad (n \geq 0)$$

$$\leq \frac{2d_n}{1 + \sqrt{1 - 2K(1 - K\Delta_n)^{-1}d_n}} \quad (n \geq 0) \qquad \text{(Moret [13], Yamamoto [30])}$$

$$\leq \frac{2d_n}{1 + \sqrt{1 - 2K(1 - Kt_n)^{-1}d_n}} \quad (n \geq 0) \qquad \text{(Yamamoto [30])}$$

$$= \frac{2d_n}{1 + \sqrt{1 - 2KB_n d_n}} \quad (n \geq 0) \qquad \text{(Yamamoto [29], [30])}$$

$$= \begin{cases} \dfrac{2d_n}{1 + \sqrt{1 - \frac{4}{\Delta}\frac{1-\theta^{2^n}}{1+\theta^{2^n}}d_n}} & (2h < 1) \\[3ex] \dfrac{2d_n}{1 + \sqrt{1 - \frac{2^n}{\eta}d_n}} & (2h = 1) \quad (n \geq 0) \end{cases} \qquad \text{(Miel [12])}$$

$$\leq \frac{t^* - t_n}{\nabla t_{n+1}}d_n \quad (n \geq 0) \qquad \text{(Yamamoto [30])}$$

$$= \frac{2d_n}{1 + \sqrt{1 - 2h_n}} \quad (n \geq 0) \qquad \text{(Döring [3])}$$

$$\leq \frac{KB_n d_{n-1}^2}{1 + \sqrt{1 - 2h_n}} \quad (n \geq 1) \qquad \text{(Döring [3])}$$

$$= \frac{t^* - t_n}{(\nabla t_n)^2}d_{n-1}^2 \quad (n \geq 1) \qquad \text{(Miel [11])}$$

$$= \begin{cases} \dfrac{1 - \theta^{2^n}}{\Delta}d_{n-1}^2 & (2h < 1) \\[3ex] \dfrac{2^{n-1}}{\eta}d_{n-1}^2 & (2h = 1) \quad (n \geq 1) \end{cases} \qquad \text{(Miel [12])}$$

$$\leq \frac{Kd_{n-1}^2}{\sqrt{1 - 2h} + \sqrt{1 - 2h + (Kd_{n-1})^2}} \quad (n \geq 1) \qquad \text{(Potra-Pták [19])}$$

$$\leq \frac{K\eta_{n-1}d_{n-1}}{\sqrt{1 - 2h} + \sqrt{1 - 2h + (K\eta_{n-1})^2}} \quad (n \geq 1) \qquad \text{(Yamamoto [29])}$$

$$= e^{-2^{n-1}\phi}d_{n-1} \quad (h^{-1} = 1 + \cosh\phi, \ \phi \geq 0) \quad (n \geq 1) \qquad \text{(Ostrowski [17])}$$

$$= \theta^{2^{n-1}}d_{n-1} \quad (n \geq 1) \qquad \text{(Gragg-Tapia [4])}$$

$$= \frac{t^* - t_n}{\nabla t_n}d_{n-1} \quad (n \geq 1) \qquad \text{(Miel [11])}$$

$$\leq t^* - t_n \quad (n \geq 0) \qquad \text{(Kantorovich [6])}$$

$$= \frac{2\eta_n}{1 + \sqrt{1 - 2h_n}} \quad (n \geq 0) \qquad \text{(Kantorovich [5])}$$

$$= \begin{cases} e^{-2^{n-1}\phi}\dfrac{\sinh\phi}{\sinh 2^{n-1}\phi}\eta & (2h < 1) \\[3ex] 2^{1-n}\eta & (2h = 1) \quad (n \geq 0) \end{cases} \qquad \text{(Ostrowski [16], [17])}$$

$$= \begin{cases} \dfrac{\Delta\theta^{2^n}}{1 - \theta^{2^n}} & (2h < 1) \\[3ex] 2^{1-n}\eta & (2h = 1) \quad (n \geq 0) \end{cases} \qquad \text{(Gragg-Tapia [4])}$$

$$= \frac{s_n}{2^n K}\left(\frac{2h}{1 + \sqrt{1 - 2h}}\right)^{2^n} \left(s_0 = 1, \ s_n = \frac{s_{n-1}^2}{2^{n-1}\sqrt{1 - 2h} + s_{n-1}(1 - \sqrt{1 - 2h})^{2^{n-1}}}\right)$$
$$(n \geq 0) \qquad \text{(Rall-Tapia [22])}$$

$$\leq \frac{1}{2^n K} \left(\frac{2h}{1 + \sqrt{1 - 2h}} \right)^{2^n} \quad (n \geq 0) \qquad \text{(Dennis [2], Tapia [26])}$$

$$\leq \frac{1}{2^{n-1}} (2h)^{2^n - 1} \eta \quad (n \geq 0) \qquad \text{(Kantorovich [4])}$$

On the other hand, it is easy to see that

$$d_n \leq \tfrac{1}{2} L_n d_{n-1}^2, \qquad L_n \leq \frac{L_{n-1}}{1 - L_{n-1} d_{n-1}}, \qquad n \geq 1.$$

Therefore, we also have from Theorem 3.2

$$\|x^* - x_n\| \leq \lambda_n \quad (n \geq 0)$$

$$\leq \frac{L_n d_{n-1}^2}{1 + \sqrt{1 - (L_n d_{n-1})^2}} \quad (n \geq 1)$$

$$\leq \frac{L_{n-1} d_{n-1}^2}{1 - L_{n-1} d_{n-1} + \sqrt{1 - 2 L_{n-1} d_{n-1}}} \quad (n \geq 1) \quad \text{(Kornstaedt [8])}$$

$$\leq \frac{L_{n-1} d_{n-1}^2}{1 - L_{n-1} d_{n-1}} \quad (n \geq 1) \qquad \text{(Lancaster [9])}$$

Furthermore, by noting that

$$L_n \leq \frac{L_0}{1 - L_0 \Delta_n}$$

and

$$d_n \leq \frac{1}{1 - L_0 \Delta_n} \|F'(x_0)^{-1} F(x_n)\| \leq \frac{L_0 d_{n-1}^2}{2(1 - L_0 \Delta_n)}, \qquad n \geq 1,$$

we have for $n \geq 0$

$$\|x^* - x_n\| \leq \lambda_n$$

$$\leq \frac{2 d_n}{1 + \sqrt{1 - 2 L_0 (1 - L_0 \Delta_n)^{-1} d_n}}$$

$$\leq \frac{2 \|F'(x_0)^{-1} F(x_n)\|}{1 - L_0 \Delta_n + \sqrt{(1 - L_0 \Delta_n)^2 - 2 L_0 \|F'(x_0)^{-1} F(x_n)\|}} \qquad \text{(Potra [18])}$$

$$\leq \frac{L_0 d_{n-1}^2}{1 - L_0 \Delta_n + \sqrt{(1 - L_0 \Delta_n)^2 - (L_0 d_{n-1})^2}} \qquad \text{(Potra [18])}$$

Similarly we have the following chart for the lower bounds for $\|x^* - x_n\|$:

$$\|x^* - x_n\| \geq \underline{\kappa}_n \geq \underline{\lambda}_n$$

$$\geq \frac{2 d_n}{1 + \sqrt{1 + 2K(1 - K\Delta_n)^{-1} d_n}} \quad (n \geq 0) \qquad \text{(Yamamoto [30])}$$

$$\geq \frac{2 d_n}{1 + \sqrt{1 + 2K(1 - K t_n)^{-1} d_n}} \quad (n \geq 0) \qquad \text{(Schmidt [25])}$$

$$= \frac{2d_n}{1 + \sqrt{1 + 2KB_n d_n}} \qquad (n \geq 0) \qquad \text{(Miel [11], Yamamoto [29])}$$

$$= \begin{cases} \dfrac{2d_n}{1 + \sqrt{1 + \frac{4}{\Delta} \cdot \frac{1 - \theta^{2^n}}{1 + \theta^{2^n}} d_n}} & (2h < 1) \\[4mm] \dfrac{2d_n}{1 + \sqrt{1 + \frac{2^n}{\eta} d_n}} & (2h = 1) \quad (n \geq 0) \end{cases} \qquad \text{(Miel [12])}$$

$$= \frac{2d_n}{1 + \sqrt{1 + 4\dfrac{t^* - t_{n+1}}{(t^* - t_n)^2} d_n}} \qquad (n \geq 0) \qquad \text{(Miel [12])}$$

$$= \frac{2d_n}{1 + \sqrt{1 + 4\dfrac{\nabla t_{n+1}}{\nabla t_n} d_n}} \qquad (n \geq 1) \qquad \text{(Miel [12])}$$

$$= \frac{2d_n}{1 + \sqrt{1 + \dfrac{2Kd_n}{\sqrt{1 - 2h + (K\eta_{n-1})^2}}}} \qquad (n \geq 1) \qquad \text{(Yamamoto [29])}$$

$$\geq \frac{2d_n}{1 + \sqrt{1 + \dfrac{2Kd_n}{\sqrt{1 - 2h + (Kd_{n-1})^2}}}} \qquad (n \geq 1) \qquad \text{(Yamamoto [29])}$$

$$= \frac{2d_n}{1 + \sqrt{1 + \dfrac{2d_n}{\sqrt{a^2 + d_{n-1}^2}}}} \qquad (a = \sqrt{1 - 2h}/K, \; n \geq 1) \qquad \text{(Yamamoto [29])}$$

$$\geq \frac{2d_n}{1 + \sqrt{1 + \dfrac{2d_n}{d_n + \sqrt{a^2 + d_n^2}}}} \qquad (n \geq 0) \qquad \text{(Potra-Pták [19])}$$

$$\geq \frac{2d_n}{1 + \sqrt{1 + 2h_n}} \qquad (n \geq 0) \qquad \text{(Gragg-Tapia [4])}$$

$$= \frac{2d_n}{1 + \sqrt{1 + \dfrac{4\theta^{2^n}}{(1 + \theta^{2^n})^2}}} \qquad (n \geq 0) \qquad \text{(Gragg-Tapia [4])}$$

We can also prove the following theorem.

Theorem 3.3. *With the notation and assumptions of Theorem 3.2, we have for* $n \geq 0$

$$\|x^* - x_{n+1}\| \leq \kappa_{n+1} \leq \kappa_n - d_n \leq \frac{t^* - t_{n+1}}{\nabla t_{n+1}} d_n$$

and

$$\|x^* - x_{n+1}\| \leq \lambda_{n+1} \leq \lambda_n - d_n \leq \frac{t^* - t_{n+1}}{\nabla t_{n+1}} d_n$$

4. Bounds for $\|x_{n+1} - x_n\|$.

Let the assumptions of Theorem 2.1 hold. Then it is easy to see that

$$d_n \leq \tfrac{1}{2}K_n d_{n-1}^2 \leq \tfrac{1}{2}L_n d_{n-1}^2 \leq \tfrac{1}{2}K(1 - K\Delta_n)^{-1} d_{n-1}^2, \qquad n \geq 1,$$

where K_n and L_n are defined in Theorem 3.2 and $\Delta_n = \|x_n - x_0\|$. Therefore, from Lemmas 2.1–2.5, we have the following chart for the estimates of d_n $(n \geq 1)$:

$$d_n \leq \tfrac{1}{2}K_n d_{n-1}^2$$

$$\leq \tfrac{1}{2}L_n d_{n-1}^2$$

$$\leq \tfrac{1}{2}K(1 - K\Delta_n)^{-1} d_{n-1}^2 \qquad \text{(Dennis [2])}$$

$$\leq \tfrac{1}{2}K(1 - K\Delta_{n-1} - K d_{n-1})^{-1} d_{n-1}^2 \qquad \text{(Moret [13])}$$

$$\leq \tfrac{1}{2}K(1 - K t_n)^{-1} d_{n-1}^2 \qquad \text{(Dennis [2])}$$

$$= \tfrac{1}{2}K B_n d_{n-1}^2 \qquad \text{(Kantorovich [5])}$$

$$= \frac{t^* - t_{n+1}}{(t^* - t_n)^2} d_{n-1}^2 \qquad \text{(Miel [12])}$$

$$= \frac{\nabla t_{n+1}}{(\nabla t_n)^2} d_{n-1}^2 \qquad \text{(Miel [11])}$$

$$= \frac{d_{n-1}^2}{2\sqrt{a^2 + \eta_{n-1}^2}} \qquad \text{(Yamamoto [29])}$$

$$\leq \frac{d_{n-1}^2}{2\sqrt{a^2 + d_{n-1}^2}} \qquad \text{(Yamamoto [29])}$$

$$\leq \frac{\nabla t_{n+1}}{\nabla t_n} d_{n-1} \qquad \text{(Miel [11])}$$

$$= \frac{\eta_n}{\eta_{n-1}} d_{n-1}$$

$$= \frac{1}{2 \cosh 2^{n-1} \phi} d_{n-1} \qquad \text{(Ostrowski [17])}$$

$$\leq \tfrac{1}{2} d_{n-1}$$

$$\leq \eta_{n-1} = \tfrac{1}{2}\nabla t_n.$$

Furthermore, for $n \geq 0$, we have

$$d_n \leq \eta_n \qquad \text{(Kantorovich [5])}$$

$$= \nabla t_{n+1} \qquad \text{(Kantorovich [6])}$$

$$= (t^* - t_{n+1}) \theta^{-2^n}$$

$$= (t^* - t_{n+1})e^{2^n \phi}$$

$$= \begin{cases} \dfrac{\sinh \phi}{\sinh 2^n \phi}\, \eta & (2h < 1) \\ 2^{-n}\eta & (2h = 1) \end{cases} \qquad \text{(Ostrowski [17])}$$

$$= \dfrac{\Delta \theta^{2^n}}{1 - \theta^{2^{n+1}}} \qquad\qquad \text{(Potra-Pták [19])}$$

Acknowledgment. The author would like to thank Professor R. K. Guy of the University of Calgary for bringing the reference [9] to his attention.

References.

[1] R. G. Bartle, "Newton's method in Banach spaces," *Proc. Amer. Math. Soc.*, **6**(1955), 827–831.

[2] J. E. Dennis, Jr., "On the Kantorovich hypothesis for Newton's method," *SIAM J. Numer. Anal.*, **6**(1969), 493–507.

[3] B. Döring, "Über das Newtonsche Näherungsverfahren," *Math. Phys. Sem.-Ber.*, **16**(1969), 27–40.

[4] W. B. Gragg and R. A. Tapia, "Optimal error bounds for the Newton-Kantorovich theorem," *SIAM J. Numer. Anal.*, **11**(1974), 10–13.

[5] L. V. Kantorovich, "On Newton's method for functional equations," *Dokl. Akad. Nauk. SSSR*, **59**(1948), 1237–1240.

[6] L. V. Kantorovich, "The majorant principle and Newton's method," *Dokl. Akad. Nauk. SSSR*, **76**(1951), 17–20.

[7] L. V. Kantorovich and G. P. Akilov, *Functional Analysis in Normed Spaces*, Pergamon Press, Oxford, 1964.

[8] H. J. Kornstaedt, "Funktionalungleichungen und Iterationsverfahren," *Aequat. Math.*, **13**(1975), 21–45.

[9] P. Lancaster, "Error analysis for the Newton-Raphson method," *Numer. Math.*, **9**(1966), 55–68.

[10] G. J. Miel, "The Kantorovich theorem with optimal error bounds," *Amer. Math. Monthly*, **86**(1979), 212–215.

[11] G. J. Miel, "Majorizing sequences and error bounds for iterative methods," *Math. Comp.*, **34**(1980), 185–202.

[12] G. J. Miel, "An updated version of the Kantorovich theorem for Newton's method," *Computing*, **27**(1981), 237–244.

[13] I. Moret, "A note on Newton type iterative methods," *Computing*, **33**(1984), 65–73.

[14] J. M. Ortega, "The Newton-Kantorovich theorem," *Amer. Math. Monthly*, **75**(1968), 658–660.

[15] J. M. Ortega and W.C. Rheinboldt, *Iterative Solution of Nonlinear Equations in Several Variables*, Academic Press, New York, 1970.

[16] A. M. Ostrowski, "La method de Newton dans les espaces de Banach," *C. R. Acad. Sci. Paris*, **27(A)**(1971), 1251–1253.

[17] A. M. Ostrowski, *Solution of Equations in Euclidean and Banach Spaces*, Academic Press, New York, 1973.

[18] F. A. Potra, "On the a posteriori error estimates for Newton's method," *Beiträge zur Numerische Mathematik*, **12**(1984), 125–138.

[19] F. A. Potra and V. Pták, "Sharp error bounds for Newton's process," *Numer. Math.*, **34**(1980), 63–72.

[20] L. B. Rall, "A note on the convergence of Newton's method," *SIAM J. Numer. Anal.*, **11**(1974), 34–36.

[21] L. B. Rall, *Computational Solution of Nonlinear Operator Equations*, Krieger, Huntington, New York, 1979.

[22] L. B. Rall and R. A. Tapia, "The Kantorovich theorem and error estimates for Newton's method," *MRC Technical Summary Report No. 1043*, University of Wisconsin, 1970.

[23] W. C. Rheinboldt, "A unified convergence theory for a class of iterative process," *SIAM J. Numer. Anal.*, **5**(1968), 42–63.

[24] J. W. Schmidt, "Regular-falsi Verfahren mit konsistenter Steigung und Majo-rantenprinzip," *Periodica Math. Hungarica*, **5**(1974), 187–193.

[25] J. W. Schmidt, "Unter Fehlerschranken für Regular-falsi-Verfahren," *Periodica Math. Hungarica*, **5**(1978), 241–247.

[26] R. A. Tapia, "The Kantorovich theorem for Newton's method," *Amer. Math. Monthly*, **78**(1971), 389–392.

[27] T. Yamamoto, "Error bounds for Newton's process derived from the Kantorovich theorem," *Japan J. Appl. Math*, **2**(1985), 285–292.

[28] T. Yamamoto, "Error bounds for Newton's iterates derived from the Kantorovich theorem," *MRC Technical Summary Report No. 2843*, University of Wisconsin, 1985; *Numer. Math.*, **48**(1986), 91–98.

[29] T. Yamamoto, "A unified derivation of several error bounds for Newton's process," *J. Comp. Appl. Math.*, **12** & **13**(1985), 179–191.

[30] T. Yamamoto, "Error bounds for Newton-like methods under Kantorovich type assumptions," *MRC Technical Summary Report No. 2846*, University of Wisconsin, 1985.

[31] T. Yamamoto, "A convergence theorem for Newton's method in Banach spaces," *MRC Technical Summary Report No. 2879*, University of Wisconsin, 1985.

[32] T. Yamamoto, "A method for finding sharp error bounds for Newton's method under the Kantorovich assumptions," to appear in *Numer. Math.*

[33] T. J. Ypma, "Affine invariant convergence results for Newton's method," *BIT*, **22**(1982), 108–118.

The author received his Ph.D. in 1968 under Y. Nakai at Hiroshima University. He has held positions at Hiroshima University and Ehime University. He is now Professor of Mathematics at Ehime University (since 1975). His interests are in numerical analysis, especially numerical algebra and nonlinear equations in Euclidean and Banach spaces.

PANEL DISCUSSION
IMPLICATIONS FOR UNDERGRADUATE
AND GRADUATE MATHEMATICS EDUCATION

SOL GARFUNKEL

Consortium for Mathematics and Its Applications
60 Lowell Street
Arlington, Massachusetts 02174

Introduction.

Participating in a conference honoring Gail Young was an important event for me. Gail has been a close friend for a number of years and it was a true pleasure to take part in demonstrating the respect and genuine affection which the mathematics community has for him. I was therefore quite pleased when I was asked to write an article summarizing the panel discussion on mathematics education. But the editors gave me an even broader charge, suggesting that I could take these few pages and write what I felt were the important educational implications coming out of the conference.

Fortunately, before sitting down to write, I listened to the audio tape of the panel discussion. My plans for the paper immediately changed as I realized the importance of what had been said and the significantly broad range of views that were covered. All thoughts of personal polemics gone, I recognized that my job was to relate those views and the spirit in which they were addressed as faithfully as possible.

The speakers, in order, were Gail Young, Sol Garfunkel, Bill Duren, and Peter Hilton. In this paper I will summarize their remarks, taking the editorial license of putting my own comments last. My only regret is the difficulty of conveying the mood of this remarkable conference to people who weren't fortunate enough to attend. The panelists and other conferees are strong-minded people with strongly held views, especially in the area of education. But the spirit of the conference, like the personality of the man it honored, was one of warmth, sincerity and cooperation.

The Panel.

Ken Gross introduced the panelists and charged us all to be freewheeling, informal, participatory, and even controversial. We were free to express points of view we felt needed to be expressed even if they did not relate directly to the talks at the conference. Our initial statements were limited to 10 minutes each.

Gail began his talk by stating the one clear message of the conference: there is *one* mathematics. The more we learn, the more clearly this fact comes through. Unfortunately, our courses and curricula are not structured to make this message clear to our students. Somehow we leave the impression that mathematical knowledge is fragmented into subdivisions. We continue this tactic even in graduate school. To reflect the underlying unity of the subject, there really should be three graduate courses: first-year graduate mathematics, second-year graduate mathematics, and third-year graduate mathematics. The great trouble is finding teachers for this course of study. Very few of us have been educated to take this broad and unified perspective.

Gail noted that each conference speaker demonstrated the unexpected ways in which methods of computation have been introduced in mathematics, both pure and applied. But the real usage of computers in college classrooms is remarkably small. For example, in 1980 only 3% of calculus classes used computers in any way at all. And yet in a very real sense, the introduction of computers in modern mathematics compares with the introduction of arabic numerals in the Middle Ages. The message is clear. It is up to us to broaden ourselves both in our mathematical and computational skills, in order to provide the instruction our students need and deserve.

Bill Duren spoke about the position of mathematicians in American society. He reminded us that mathematicians as a group have never truly held an important place. We have not been the movers and shakers or shapers of society. Our main source of income remains dependent on academic jobs which are threatened by recession and demography as well as a public resentment of costs. Yet, if this conference has demonstrated anything, it is the importance of mathematics. If mathematics is so clearly important, then surely mathematicians can and should be. Certainly there will be new jobs for mathematicians outside of academia. But if all we do is create and fill new jobs, we will not have truly redefined our role in American society.

We must become masters of our own fate as opposed to hired help. But in order to accomplish this we must change our self image and educate future mathematicians differently. There is a natural comparison with the legal profession. Legal education is geared to a solid general education. Lawyers are also leaders in politics and industry and are visible in public and private life. They see their training as a base for work outside their own field. This can and should be the way we train and advise our students. We have a responsibility to raise our students' sights away from "a job" to striking out on their own—to creating their own place in society.

Peter Hilton spoke of the relationship between mathematics and other disciplines, specifically the increased applicability of mathematics. In a real sense, the mathematics we now teach to students in other disciplines has dramatically increased in importance. In earlier years, the specific mathematical content of service courses was less important. But now we need to be more careful. The undergraduate curriculum should be redesigned with applications included as motivation and integrated throughout. The computer must be brought in organically, with more emphasis on approximation theory and numerical analysis.

We have made artificial distinctions between pure and applied mathematics. But in introducing applications we must take care to avoid oversimplification. Applications must be real—to us and to our students. But accepting this view, we have to ask what we can do, given the present state of high school mathematics education. There are no easy answers, but there is one inescapable conclusion. The college mathematics community, researchers and educators, must dedicate itself to the improvement of pre-college instruction.

Concluding Remarks.

As I warned, invoking editorial license, I'll conclude this paper with my own remarks. I began my talk at the conference by relating a statistic which I have always found remarkable. College students on the average take only one semester of mathematics. I realize that this can be misleading. But consider how our educational philosophy and curriculum decisions might change if we acted on this information. If you knew you had a college student in a mathematics class for only a semester, what would you teach?

I've asked myself and my colleagues that question many times over the past few years. The answer forms the basis for much of the work produced by COMAP. But what is intriguing is that almost all mathematicians give an answer which relates mathematics and its applications to the broad context of human endeavor. Yet, this is precisely what we fail to do in our precollege, college, and graduate instruction.

We often decry the fact that people (including politicians and budget makers) don't understand what we do and the importance and centrality of mathematics. We complain about being mistreated and misunderstood. But we must shoulder the bulk of the blame. After all, there is one arena in which we have total control—our own classroom. Yet, here in our own domain we do precious little to provide the rich context in which mathematics operates. And it is the students in those classes who are the future decision makers we wish to influence.

This conference celebrated the strength, growth, and unity of mathematics. It vividly demonstrated the power and pervasiveness of our subject. It is now up to us to carry that message into the classroom and throughout our society.

Sol Garfunkel was trained as a mathematical logician. He is presently the executive director of COMAP (Consortium for Mathematics and Its Applications). He is project director and program host for the twenty-six one-hour public television series "For All Practical Purposes."

William L. Duren, Jr., graduated from Tulane University in 1926. He received his Ph.D. at the University of Chicago under G. A. Bliss in calculus of variations and was assistant to Marston Morse at the Institute for Advanced Study before returning to Tulane University. He remained at Tulane from 1931 to 1955, when as Chairman he founded the graduate program. From 1955 to 1976 he served at the University of Virginia as University Professor and Dean of Arts and Sciences. At Virginia he founded the graduate program in applied mathematics and computer science. Dr. Duren was the first Program Director at NSF. He has been President of the MAA, and is credited with founding CUPM. His honors include phi beta kappa, an honorary degree from Tulane, and the MAA Distinguished Service Award. Throughout his career he took steps to improve opportunities in higher education for black students and professors.

EPILOGUE

by

GAIL YOUNG

I am extremely fortunate, beyond my desserts, to have my seventieth birthday chosen for such a spectacular conference. Those who have read through the papers before coming to these last remarks will agree to the adjective "spectacular." But beyond my desire to express my gratitude, I want to say something about conclusions I draw from the Conference.

I believe that we are participating in a revolution in mathematics as profound as the introduction of arabic numerals in Europe, or the invention of the calculus. Those earlier revolutions had common features: hard problems became easy, and solvable not only by an intellectual elite but by a multitude of people without special mathematical talent; problems arose that had not been previously visualized, and their solutions changed the entire level of the field.

In our case, the computer is the new tool. It is not only the immense numerical power of the computer, but also its power of symbolic manipulation that will produce the change. To consider only the effect that I can see most clearly: what will happen to our undergraduate curriculum when, for example, every manipulative problem in differential equations or advanced calculus can be "solved" on a personal computer by a student who could not possibly duplicate the solution by hand? It took a century for arabic numerals to replace roman; the change in our time is moving much faster. I was on the faculty of the University of Michigan in the 1950's, and we were proud of our statistical computing laboratory, consisting of some twenty Friden and Marchant calculators on long tables. From that to the calculations in, e.g., Dick Ewing's paper, seems to me incredible; it brings the science fiction of the 1950's to life. These calculations will soon be outstripped with the next generation of computers. In the universities and colleges we are not moving fast enough to keep up, a charge that includes me, for I have never run a program in my life.

The summer before this Conference, Lipman Bers wrote me that when we began, I was a topologist and he was some sort of applied mathematician. "Now we both know there is only one mathematics." That is true, and this Conference documents

213

it. Despite efforts (e.g., Bourbaki) to provide a common basis for mathematics, most of us still believe in specialties. But, to take someone outside this Conference, exactly what is Atiyah's field? It can be only mathematics. I am sure that there are students now in high school who will receive their Ph.D.'s in mathematics around 2000 A.D. whose view of our subject will be immensely broader than, for example, mine, or any but the greatest of my generation.

Peter Hilton mentioned in the Preface that soon after the Conference my wife, Irene, died. She and I both knew that her death would be only a matter of weeks. She was tremendously happy during the Conference, and proud of me. Her happiness enabled me to enjoy it also, something I had not thought possible in the months before the Conference. Our nearly 18 years of marriage were wonderful for us both. I will always miss her.

There has been a certain overemphasis in this volume on my retiring. I think I have only stopped teaching.

There are so many people to thank that I will not really attempt to name them. But of course it includes the three organizers, the speakers and the participants. I will mention specifically the presence of Bill Duren. It is rare that at a seventieth birthday conference an undergraduate teacher of the honoree is present. I have leaned on his wisdom and judgment all my career. I am particularly pleased that so many old friends (and new ones) were speakers. Perhaps that is partly responsible for the remarkable clarity of the talks; knowing me, they made sure I could follow.